Conference Proceedings of the Society for Experimental Mechanics Series

Series Editor

Kristin B. Zimmerman, Ph.D.
Society for Experimental Mechanics, Inc.,
Bethel, CT, USA

More information about this series at http://www.springer.com/series/8922

Martha Grady • Majid Minary • La Vern Starman • Jenny Hay
Jacob Notbohm
Editors

Mechanics of Biological Systems & Micro-and Nanomechanics, Volume 4

Proceedings of the 2018 Annual Conference on Experimental and Applied Mechanics

Springer

Editors
Martha Grady
Department of Mechanical Engineering
College of Engineering
University of Kentucky
Lexington, KY, USA

La Vern Starman
Air Force Research Lab
Wright-Patterson, OH, USA

Jacob Notbohm
University of Wisconsin-Madison
Madison, WI, USA

Majid Minary
Department of Mechanical Engineering
University of Texas at Dallas
Richardson, TX, USA

Jenny Hay
Nanomechanics, Inc
Oak Ridge, TN, USA

ISSN 2191-5644 ISSN 2191-5652 (electronic)
Conference Proceedings of the Society for Experimental Mechanics Series
ISBN 978-3-030-06966-7 ISBN 978-3-319-95062-4 (eBook)
https://doi.org/10.1007/978-3-319-95062-4

This Springer imprint is published by the registered company Springer Nature Switzerland AG
The registered company address is: Gewerbestrasse 11, 6330 Cham, Switzerland

Preface

Mechanics of Biological Systems & Micro- and Nanomechanics represents one of eight volumes of technical papers presented at the 2018 SEM Annual Conference and Exposition on Experimental and Applied Mechanics organized by the Society for Experimental Mechanics and held in Greenville, SC, from June 4 to 7, 2018. The complete proceedings also include volumes on *Dynamic Behavior of Materials; Challenges in Mechanics of Time-Dependent Materials; Advancement of Optical Methods & Digital Image Correlation in Experimental Mechanics; Mechanics of Composite, Hybrid and Multifunctional Materials; Fracture, Fatigue, Failure and Damage Evolution; Residual Stress, Thermomechanics & Infrared Imaging, Hybrid Techniques and Inverse Problems; and Mechanics of Additive and Advanced Manufacturing.*

Each collection presents early findings from experimental and computational investigations on an important area within experimental mechanics, the mechanics of biological systems, micro- and nanomechanics, and other experimental and applied mechanics such as education and research in progress.

The Biological Systems and Materials segment of this volume summarizes the exchange of ideas and information among scientists and engineers involved in the research and analysis of how mechanical loads interact with the structure, properties, and function of living organisms and their tissues. The scope includes experimental, imaging, numerical, and mathematical techniques and tools spanning various length and time scales. Establishing this symposium at the Annual Meeting of the Society for Experimental Mechanics provides a venue where state-of-the-art experimental methods can be leveraged in the study of biological and bio-inspired materials, traumatic brain injury, cell mechanics, and biomechanics in general. A major goal of the symposium was for participants to collaborate in the asking of fundamental questions and the development of new techniques to address bio-inspired problems in society, human health, and the natural world. The organizers would like to thank all the speakers and staff at SEM for enabling a successful program.

The Micro- and Nanomechanics segment of this volume focuses on specialized scientific areas that involve miniaturizing conventional scale components and systems to take advantage of reduced size and weight and/or enhanced performance or novel functionality. These fields also encompass the application of principles ranging from the micron scale down to individual atoms. Sometimes these principles borrow from conventional scale laws but often involve new physical and/or chemical phenomena that require new behavioral laws and impart new properties to exploit. Studying how mechanical loads interact with components of these scales is important in developing new applications, as well as assessing their reliability and functionality. Establishing this symposium at the Annual Meeting of the Society for Experimental Mechanics provides a venue where state-of-the-art experimental methods can be leveraged in these endeavors.

The 2018 Symposium is the nineteenth in the series and addresses pertinent issues relating to design, analysis, fabrication, testing, optimization, and applications of micro- and nanomechanics, especially as these issues relate to experimental mechanics of microscale and nanoscale structures.

It is with deep gratitude that we thank the organizing committee, session chairs, authors and keynote speakers, participants, and SEM staff for making the 19th *International Symposium on Micro- and Nanomechanics (ISMAN)* a valuable and unforgettable experience.

Lexington, KY, USA	Martha Grady
Richardson, TX, USA	Majid Minary
Madison, WI, USA	Jacob Notbohm
Wright-Patterson AFB, OH, USA	La Vern Starman
Oak Ridge, TN, USA	Jenny Hay

Contents

Contributors

Stephen Alexander SURVICE Engineering Company, Belcamp, MD, USA

Brian E. Allgeier University of Louisville, Louisville, KY, USA

Dilip K. Banerjee National Institute of Standards and Technology, Gaithersburg, MD, USA

C. Barile Dipartimento di Meccanica, Matematica e Management (DMMM), Politecnico di Bari, Bari, Italy

Thomas A. Berfield University of Louisville, Louisville, KY, USA

Sharada Bhavanam Aerospace Engineering Program, Department of Mechanical Engineering, Worcester Polytechnic Institute, Worcester, MA, USA

Daniel Biggs Division of Engineering and Applied Science, California Institute of Technology, Pasadena, CA, USA

James D. Boyd Department of Mechanical Engineering, College of Engineering, University of Kentucky, Lexington, KY, USA

Natasha Bradley U.S. Army Research Laboratory, Vehicle Technology Directorate, Aberdeen Proving Ground, MD, USA

Michael C. Bustamante Department of Mechanical Engineering, University of Waterloo, Waterloo, ON, Canada

C. Casavola Dipartimento di Meccanica, Matematica e Management (DMMM), Politecnico di Bari, Bari, Italy

A. Cazzato Dipartimento di Meccanica, Matematica e Management (DMMM), Politecnico di Bari, Bari, Italy

John Chen U.S. Army Research Laboratory, Vehicle Technology Directorate, Aberdeen Proving Ground, MD, USA

Duane S. Cronin Mechanical and Mechatronics Engineering, University of Waterloo, Waterloo, ON, Canada

Samantha Daly Department of Mechanical Engineering, University of California, Santa Barbara, Santa Barbara, CA, USA

Maarten P. de Boer Department of Mechanical Engineering, Carnegie Mellon University, Pittsburgh, PA, USA

Masoud Derakhshani University of Louisville, Louisville, KY, USA

Sarah Dooley Sensors Directorate, Air Force Research Laboratory, Wright-Patterson AFB, OH, USA

Rong Z. Gan School of Aerospace and Mechanical Engineering, University of Oklahoma, Norman, OK, USA

Martha Grady Department of Mechanical Engineering, College of Engineering, University of Kentucky, Lexington, KY, USA

C. Allan Gunnarsson U.S. Army Research Laboratory, Aberdeen Proving Ground, MD, USA

Mulugeta Haile U.S. Army Research Laboratory, Vehicle Technology Directorate, Aberdeen Proving Ground, MD, USA

H. J. Hall Sensors Directorate, Air Force Research Laboratory, Wright-Patterson AFB, OH, USA

Christopher Hsu U.S. Army Research Laboratory, Vehicle Technology Directorate, Aberdeen Proving Ground, MD, USA

Mark A. Iadicola National Institute of Standards and Technology, Gaithersburg, MD, USA

G. M. Kamath Department of Aerospace Engineering, Indian Institute of Technology, Kanpur, India

Sudhir Kamle Department of Aerospace Engineering, Indian Institute of Technology, Kanpur, India

Nikhil Karanjgaokar Aerospace Engineering Program, Department of Mechanical Engineering, Worcester Polytechnic Institute, Worcester, MA, USA

Joseph Kerwin Department of Mechanical Engineering, Michigan State University, East Lansing, MI, USA

David Kumar Department of Aerospace Engineering, Indian Institute of Technology, Kanpur, India

R. A. Lake Air Force Institute of Technology, Wright-Patterson AFB, OH, USA

Junfeng Liang Department of Mechanical Engineering, University of Texas at Dallas, Richardson, TX, USA

W. E. Liang Department of Mechanical Engineering, National Kaohsiung University of Science and Technology, Kaohsiung, Taiwan

S. Y. Lin Department of Mechanical Engineering, National Kaohsiung University of Science and Technology, Kaohsiung, Taiwan

Marissa Linne Department of Materials Science and Engineering, University of Michigan, Ann Arbor, MI, USA

Hanwei Liu Division of Chemistry and Chemical Engineering, California Institute of Technology, Pasadena, CA, USA

Hongbing Lu Department of Mechanical Engineering, University of Texas at Dallas, Richardson, TX, USA

Daniel J. Magagnosc US Army Research Laboratory, Aberdeen Proving Ground, MD, USA

Faezeh Masoomi Department of Mechanical Engineering, Michigan State University, East Lansing, MI, USA

Ricardo Mejia-Alvarez Department of Mechanical Engineering, Michigan State University, East Lansing, MI, USA

P. M. Mohite Department of Aerospace Engineering, Indian Institute of Technology, Kanpur, India

Changho Oh Department of Mechanical Engineering, Carnegie Mellon University, Pittsburgh, PA, USA

Karin Rafaels U.S. Army Research Laboratory, Aberdeen Proving Ground, MD, USA

Guruswami Ravichandran Division of Engineering and Applied Science, California Institute of Technology, Pasadena, CA, USA

Stuart C. Ross Department of Mechanical Engineering, College of Engineering, University of Kentucky, Lexington, KY, USA

Michael Sangid Department of Aeronautics and Astronautics, Purdue University, West Lafayette, IN, USA

Brian E. Schuster US Army Research Laboratory, Aberdeen Proving Ground, MD, USA

Nelson Sepúlveda Department of Electrical and Computer Engineering, Michigan State University, East Lansing, MI, USA

La Vern Starman Sensors Directorate, Air Force Research Laboratory, Wright-Patterson AFB, OH, USA

Michaelann Tartis Department of Chemical Engineering, New Mexico Institute of Mining and Technology, Socorro, NM, USA

David Tirrell Division of Chemistry and Chemical Engineering, California Institute of Technology, Pasadena, CA, USA

D. Torres Sensors Directorate, Air Force Research Laboratory, Wright-Patterson AFB, OH, USA

L. Tsai Department of Mechanical Engineering, National Kaohsiung University of Science and Technology, Kaohsiung, Taiwan

Ajey Venkataraman Department of Aeronautics and Astronautics, Purdue University, West Lafayette, IN, USA

Suhas Vidhate Department of Mechanical Engineering, Michigan State University, East Lansing, MI, USA

Tim Walter U.S. Army Research Laboratory, Aberdeen Proving Ground, MD, USA

J. P. Walton Sensors Directorate, Air Force Research Laboratory, Wright-Patterson AFB, OH, USA

Brock Watson Mechanical and Mechatronics Engineering, University of Waterloo, Waterloo, ON, Canada

Tusit Weerasooriya U.S. Army Research Laboratory, Aberdeen Proving Ground, MD, USA

Adam M. Willis Department of Mechanical Engineering, Michigan State University, East Lansing, MI, USA

Department of Neurology, San Antonio Military Medical Center, Fort Sam Houston, TX, USA

Michael J. Worswick Mechanical and Mechatronics Engineering, University of Waterloo, Waterloo, ON, Canada

Chapter 1
Investigation of Deformation Mechanisms in High Purity Columnar Aluminum

Marissa Linne, Ajey Venkataraman, Michael Sangid, and Samantha Daly

Abstract Predictive modeling of structural metallic alloys, including the ability to tailor materials for energy-efficient use in advanced applications, is hindered by a limited understanding of the interactions between deformation mechanisms and microstructural features. Deformation mechanisms such as dislocation slip and grain boundary sliding are largely treated as independent events, irrespective of indications that they may be synergistic. Limited experimental data makes it difficult to accurately predict damage localization. In this work, new experimental approaches are used to characterize mechanism-mechanism relationships and the effect of local microstructure in oligocrystalline, high purity aluminum. Using SEM-DIC and EBSD, high-resolution displacement fields and grain orientations are captured and analyzed to identify deformation trends with respect to local microstructure. Results are integrated with a parallel crystal plasticity model to improve predictions of damage-inducing strain localization, with the ultimate goal of improving the capability to develop materials with greater resistance to failure.

Keywords Grain boundary sliding · Slip transmission · Crystal plasticity · DIC · EBSD

1.1 Overview

Dislocation slip and grain boundary sliding (GBS) are deformation mechanisms that are critically important to plastic deformation behavior. However, little is known about the interactions between these mechanisms, and this lack of information has heavily limited the ability to accurately model and predict the mechanical behavior of structural metallic materials. It is hypothesized that a synergistic relationship exists between dislocation slip and GBS. However, due to a lack of experimental information, modeling of structural metals generally treats these deformation mechanisms as completely separate events, which is likely a source of significant error in our ability to predict damage initiation and propagation in these materials. The relationships between these deformation modes—whether collaborative or competitive, simultaneous or sequential—remain unclear and supported by limited experimental data. These ambiguities lead to inaccuracies in material modeling, and consequently a failure to optimize material design. This experimental investigation is aimed towards improving predictions of the initiation and propagation of damage in materials, and ultimately towards improving the capability to develop structural metallic alloys with greater resistance to failure.

The specific aims of this research are to (1) quantitatively characterize deformation behavior across grain boundaries across large (mm-scale) fields of view, with very high spatial resolution, providing local detail around the boundary as well as information on long-range interactions in the material; (2) explore the temperature dependence of these mechanisms to further probe mechanism interaction for high temperature applications; (3) achieve the above two aims with the development of new experimental methodologies for mechanical testing and acquiring full-field deformation and microstructure data; (4) use full-field deformation and microstructure data to inform a parallel crystal plasticity model (Sangid Group) and to provide statistical insight for improving predictions of damage-inducing strain localization.

M. Linne (✉)
Department of Materials Science and Engineering, University of Michigan, Ann Arbor, MI, USA
e-mail: marlinne@umich.edu

A. Venkataraman · M. Sangid
Department of Aeronautics and Astronautics, Purdue University, West Lafayette, IN, USA

S. Daly
Department of Mechanical Engineering, University of California, Santa Barbara, Santa Barbara, CA, USA

M. Grady et al. (eds.), *Mechanics of Biological Systems & Micro-and Nanomechanics, Volume 4*, Conference Proceedings of the Society for Experimental Mechanics Series, https://doi.org/10.1007/978-3-319-95062-4_1

1.2 Experimental Methods

These goals are accomplished through the measurement of microscale strain fields resulting from uniaxial tensile loading. High purity, oligocrystalline Al specimens were loaded in uniaxial tension, and full-field strain and grain orientation information were collected using in-SEM digital image correlation (SEM-DIC) and electron backscatter diffraction (EBSD), respectively. Deformation data was collected at both room (24 °C) and elevated (190 °C) temperatures to investigate the effects of enhanced diffusion on deformation mechanism relationships. During the room temperature test samples were strained in-situ, and during the elevated temperature test samples were strained ex-situ with deformation of plastic strains after unloading at axial strain intervals. Deformation trends with respect to microstructural features were identified through analysis of the information-rich strain and microstructure data arrays.

Oligocrystalline, high purity (99.999%) aluminum was selected as the test specimen for the following reasons: (1) Al is a material of interest for light-weighting structural applications; (2) it has a high stacking fault energy, so twinning does not obscure GBS-slip interactions during deformation; and (3) through-thickness grain structure allows the microstructure to be characterized by EBSD without minimal sub-surface ambiguity. Due to their high purity, oligocrystalline microstructure and 1×5 mm gage size, these samples are extremely delicate but allow collection of a unique and highly modellable full-gage data set.

Specimen were cut from high purity sheets with thicknesses of either 0.25 or 0.5 mm, by wire electro-discharge machining. To achieve a through-thickness grain structure, all specimens were annealed at 550 °C in an argon atmosphere for 6 h. Both sides of the aluminum specimen were polished to a mirror finish using mechanical polishing and full-gage grain orientation maps were acquired from the front and back gage surfaces using EBSD. One side of the specimen was patterned for SEM-DIC with 300 nm diameter gold nanoparticles (Sigma-Aldrich), using a surface silanization technique similar to that described by Kammers and Daly [1].

Uniaxial tensile tests were performed at both room (24 °C) and elevated (190 °C) temperatures. It is expected that at this elevated temperature, enhanced diffusion increases the relative contribution of GBS to the sample deformation, compared to the room temperature case. Comparing deformation behavior between these two cases provides insight into GBS-slip mechanisms dependencies. In both tests, specimen straining is performed using a tension-compression module (Kammrath and Weiss) and SEM-DIC images are collected in an array that encompasses the entire specimen gage at progressive strain increments using a Teneo field-emission SEM (FEI Company). In the room temperature case, these images are captured *in-situ*. In the elevated temperature case, the specimen is heated and strained in atmosphere and SEM-DIC images are collected *ex-situ* after unloading and cooling. The room temperature strain fields reflect combined elastic and plastic deformation, and the elevated temperature test strain fields reflect plastic deformation. SEM-DIC images were correlated using commercial software (Correlated Solutions). SEM distortions were corrected and strain/displacement fields were stitched using custom-developed MATLAB codes.

1.3 Results and Future Work

The resulting strain fields and grain orientation maps comprise an information-rich data set. Each strain field has $\sim 10^6$ data points, each point associated with several hundred points of local microstructure and deformation information, all of which evolve in time. An example of a full-gage strain field, obtained from room temperature *in-situ* straining, overlaid with EBSD-identified grain boundaries, is seen in Fig. 1.1. The inset of Fig. 1.1 (upper-right) highlights the high spatial resolution of the strain data; individual slip traces and slip transmission events are resolved. With grain orientation data, the slip systems involved in slip transmission from Grain 1 to Grain 2 can be identified. A statistical analysis using a custom-developed MATLAB code is used to identify instances of slip transmission (such as that shown in Fig. 1.1), and GBS and correlate to underlying microstructure characteristics. In addition to statistical analysis, these experimental data are used to inform a GBS-sensitive CPFE model (Sangid Group).

Deformation trends discerned with these experimental and computational approaches will provide insight into the currently limited understanding of plastic deformation and deformation mechanisms relationships. This information will help to improve capabilities to model and predict damage-inducing strain localization and design materials structural components with greater resistance to failure.

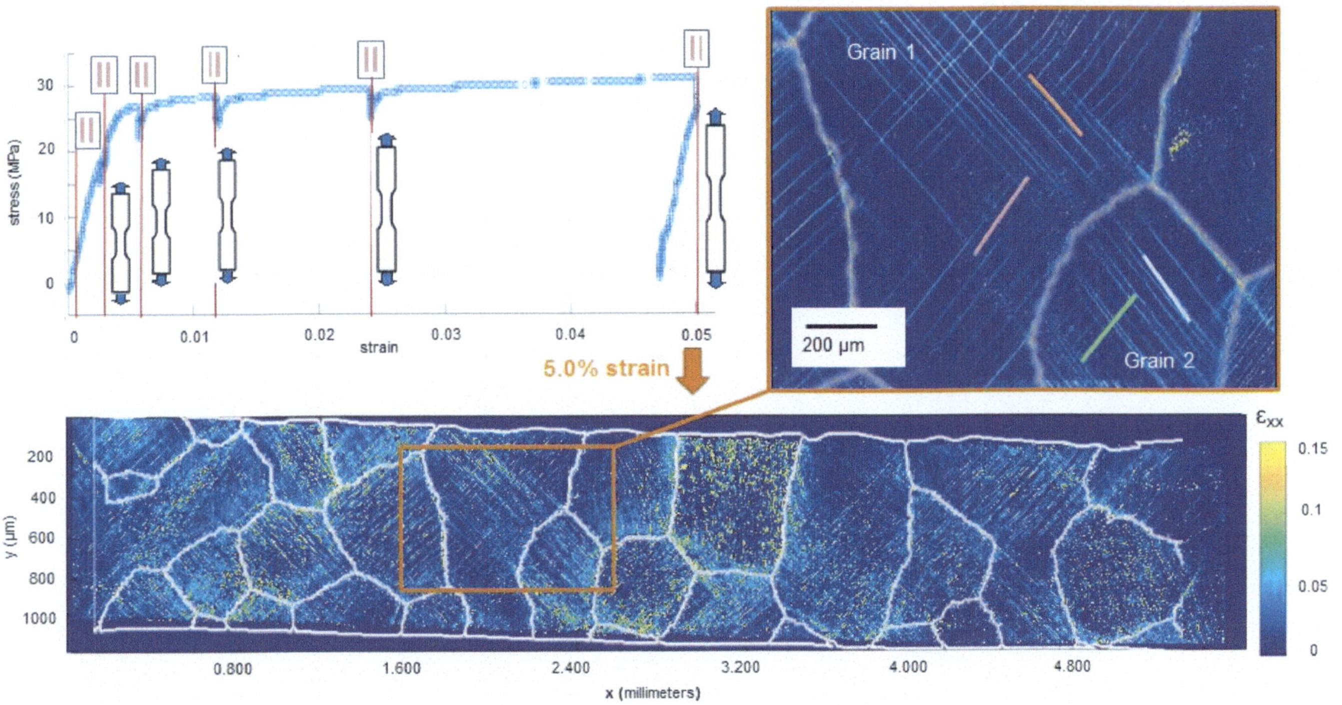

Fig. 1.1 An example data set obtained with the room temperature *in-situ* experimental approach: at each strain increment, marked on the stress-strain curve (top left), a full-gage strain field is measured (below) with high-resolution of the data set highlighted in the inset (top right)

Reference

1. Kammers, A., Daly, S.: Self-assembled nanoparticles surface patterning for improved digital image correlation in scanning electron microscope. Exp. Mech. **53**(8), 1333–1341 (2013)

Marissa Linne is a graduate student in materials science and engineering at University of Michigan.

Chapter 2
Dynamic Shear Response of Soft Tissue Materials

W. E. Liang, S. Y. Lin, and L. Tsai

Abstract During exercise, an athlete's knee could be tested under heavy loading repeatedly. In this study, dynamic shear tests of knee soft tissue under high strain rate were performed. Porcine knee tissues were prepared as the test sample. The dynamic strength of meniscus and cartilage tissues were examined at six different locations and orientations. Modified Hopkinson bar were utilized to perform the experiments. Modifications were made on incident bar and transmission bar to make sure that specimens receive force evenly and for easier shear test process. The research finding will assist the future researches on knee dynamics and soft tissues restoration.

Keywords Cartilage · Meniscus · Dynamic experiments · Shear experiments · Biomaterials

2.1 Introduction

As modern medical care and treatments advancing, the average human life expectancy becomes longer [1]. However, longer average human life expectancy may leads to aging society which is an important issue for many developed countries. Along with the growth of the age, osteoarthritis becomes a common disease [2]. Athletes also frequently injured their knees which is the main body weight support and the largest joint in human body. In 2012, Benedikt compared the Knee tissue from human to different animal, and found out that lateral meniscus size and shape of sheep, goats and porcines are similar to human's. Medial meniscus size and shape of sheep and goats are the most similar to human's [3]. Wenbo Zhu used fibrous tissue cut from meniscus layer of cow meniscus tissue from different directions (radial, axial and circumferential) to do fatigue circular shear test [4]. However, high strain rate dynamic shear and puncture research are rarely performed.

2.1.1 Articular Tissues

Articular tissues are movable joint connecting the femur and tibia. The main components are the tibia, femur and patella. The four main ligaments provide overall stability and protection. One of the cartilage tissue between the tibia and femur is the meniscus, which can avoid the direct friction of the bones, reduce the impact and provide shock absorption function. Knee injuries are mostly due to compression, stretching, friction, shearing and other factors. Knee joint contains two fibrous cartilage attached to the surface of the tibia and femur [5] and medial and lateral meniscus providing surface lubrication and distributing the impact evenly in the knee joint to increase stability [6, 7]. When the meniscus is damaged and the joint surface is worn, Inflammation of the knee joint may happen [8]. In recent years, there are more and more research on osteoarthritis (OA) especially relate to research of knee arthritis. Some studies have pointed out that the reason why knee-related diseases occur in the elderly is not only because of metabolic problems but mostly because of that the knee is inherently damaged and without regularly examined, resulting in more and more severe knee damage [9]. In some animal knee tests, it has been demonstrated that shear stress can cause significant damage to soft tissue of the knee [10–13]. The result shows that the shear resistance of samples of meniscus vertical fiber orientation in the middle section is lowest [14]. In this study, dynamic shear tests were performed on the medial and lateral menisci and the cartilages are medial and lateral to observe the dynamic shear and puncture properties of soft tissue of porcine knee under high strain rate.

W. E. Liang · S. Y. Lin · L. Tsai (✉)
Department of Mechanical Engineering, National Kaohsiung University of Science and Technology, Kaohsiung, Taiwan
e-mail: 1105303143@gm.kuas.edu.tw; liren@kuas.edu.tw

M. Grady et al. (eds.), *Mechanics of Biological Systems & Micro-and Nanomechanics, Volume 4*, Conference Proceedings of the Society for Experimental Mechanics Series, https://doi.org/10.1007/978-3-319-95062-4_2

2.1.2 *Meniscus Tear*

There are a lot of knee injuries with meniscus damage. There is sufficient micro-vascular in the most peripheral area of meniscus, so it provide nutrients to self-repair in that area. The meniscus tissue near the inside does not have the vascular to provide enough nutrients to self-repair; as a result, the internal meniscus injury often requires surgery to recover. If the pain is ignored, the meniscus damaging will get worse.

2.2 Experiment Set Up

2.2.1 *Split Hopkinson Pressure Bar*

Hopkinson is the first person who use impulse wave transmission to test the dynamic properties of material. So far, Hopkinson bar have different types, no matter which kind of Hopkinson bar, it is considered an effective and economical strain rate of 1000–10,000 s^{-1} testing method of material. This experiment used a modified version of Hopkinson bar, Split Hopkinson Pressure Bar (SHPB) or Kolsky Bar. Structurally, it is divided into Incident bar and transmitted bar, and specimen is hold between two bars. We change the bar end faces, so we can proceed puncture experiment without fixture. As shown in Fig. 2.1.

2.2.2 *Specimen Preparation*

This study utilized dynamic shearing and puncture on meniscus and cartilage to discuss dynamic shear response from join soft tissues. Human tissue specimens are difficult to access, so porcine meniscus and cartilage were used as replacement. Meniscus is constituted by three different fiber layer. The outermost layer is constituted by irregular direction fibers, and the surface of the fibers is smooth and directly contacts with the cartilage, middle section of meniscus is irregularly arranged fibers, inner layer is mostly constituted by curved orientations fiber and some radial fibers [15, 16]. We take sample from two positions on meniscus (axial and radial). Meniscus and cartilage also are divided into three sections (front, middle, back). Discuss the puncture responses difference among different sections. Specimen diameter is 5 mm, thickness is 1 mm. As shown in Fig. 2.2.

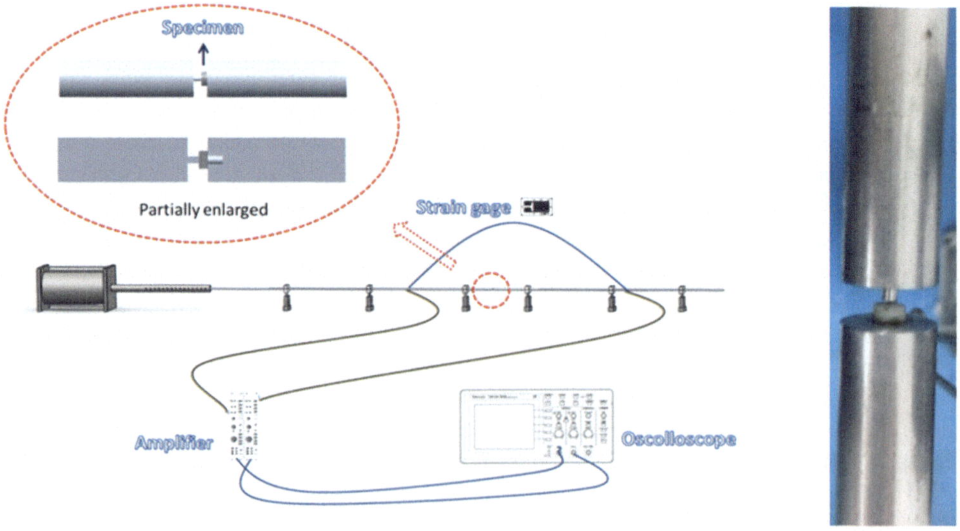

(a) (b)

Fig. 2.1 (**a**) Modified Split-Hopkinson pressure bar; (**b**) specimen hold in the bar

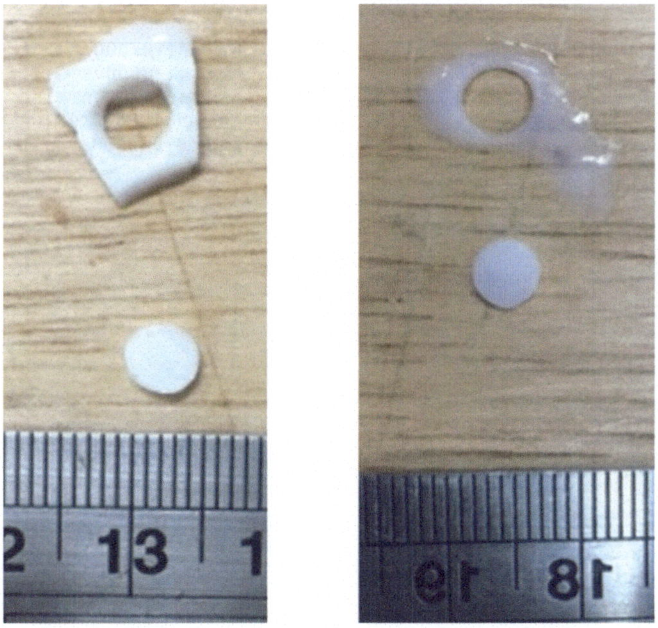

(a) (b)

Fig. 2.2 The actual specimen. (**a**) Meniscus; (**b**) cartilage tissue

2.3 Results and Discussion

Currently, there is no experiment on puncture and shearing of knee soft tissue under high strain. Therefore, this experiment mainly explores the puncture and shear properties of porcine knee soft tissue at super high strain rate after soaking in human body fluid at 37 °C. Split Hopkinson Pressure Bar (SHPB) was utilized to perform strain puncture experiments. Lateral cartilage is more likely to come into contact with most menisci because of its location. The medial meniscus is subjected to more force and injury than the lateral meniscus. We used lateral cartilage and medial meniscus as experimental specimens in this study. Experimental specimens were taken from menisci tissues of different fiber directions (radial, axial, and circumferential). The specimen was soaked in a 37° SBF solution then punctured immediately. Experimental axial shear test strain rate range are between 6000 and 8500 s^{-1}, the ultimate shear stress is 0.68–1.03 MPa (average: 0.88 MPa). As shown in Fig. 2.3, the shear modulus is 0.66–2.82 MPa (average: 1.23 MPa). The radial range is from 6500 to 9000 s^{-1}, and the ultimate shear stress is 0.08–0.534 MPa (average: 0.3 MPa). The shear modulus is 0.08–0.47 MPa (average: 0.28 MPa) as shown in Fig. 2.4. The range of the circumferential direction is from 4000 to 5500 s^{-1}, the ultimate shear stress is 0.021–0.15 MPa (average: 0.072 MPa), and the shear modulus is 0.007–0.067 MPa (average: 0.024 MPa) as shown in Fig. 2.5, Table 3 shows the data. The experimental strain rate of cartilage ranged from 6500 to 8500 s^{-1}, the ultimate shear stress was 0.382–0.916 MPa (average: 0.671 MPa) and the shear modulus was 0.29–0.801 MPa (average: 0.52 MPa) as shown in Figs. 2.6 and 2.7.

2.4 Conclusions

1. From the stress-strain diagram in Figs. 2.5, 2.6, and 2.7, it is clear that the meniscus has significantly different maximum shear strength and shear modulus in different specimens. In the meniscus axial specimen ultimate shear stress average of 0.88 MPa, the average shear modulus of 1.23 MPa is the highest among the three samples. The average ultimate shear stress of radial specimens is 0.3 MPa and the average shear modulus is 0.28 MPa, is the middle in three samples. The average shear stress of the sample in the circumferential direction is 0.072 MPa, the average of the shear modulus is 0.024 MPa, is the minimum of the three samples. In this study, we choose inner layer, because the orientation of the inner layer of the meniscus mostly consists of a combination of curved orientation and a small portion of the radial orientation.

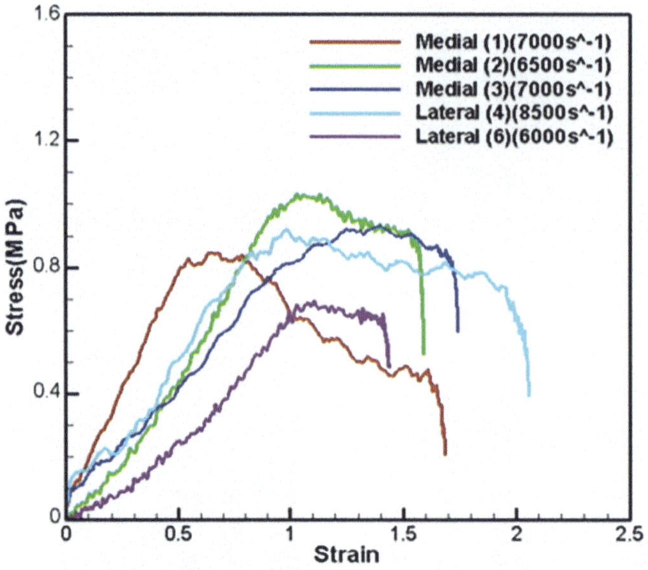

Fig. 2.3 Dynamic shear meniscus axial shear stress-strain curve

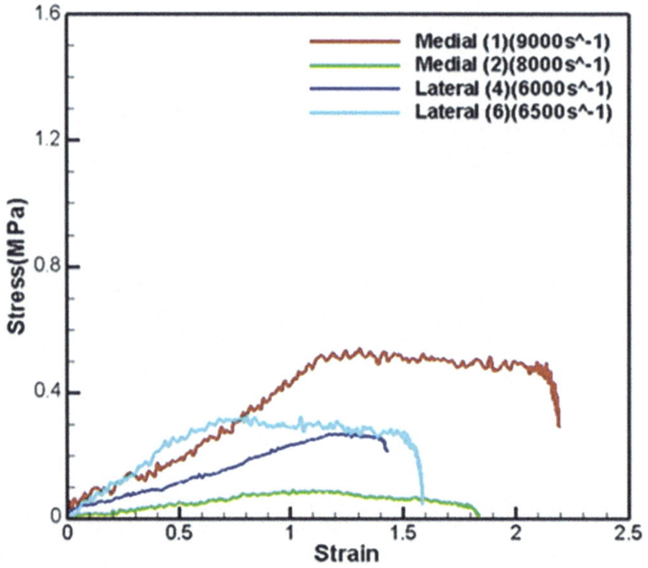

Fig. 2.4 Dynamic shear meniscus radial cut stress-strain curve

2. The experiment in accordance with the different positions as shown in Fig. 2.8 for the meniscus divided by different ways to cut out the different parts of the figure represented by the largest shear stress and modulus trend in the comparison. In Fig. 2.8a it is hard to observe a regular trend. Figure 2.8d shows a comparison of porcine cartilage tissue. From the figure, it can be observed that the shear strength of the distal part (−6) tends to be higher than that of the other parts, and it can be confirmed that the cartilage on the lateral front has relatively good anti-shearing ability.

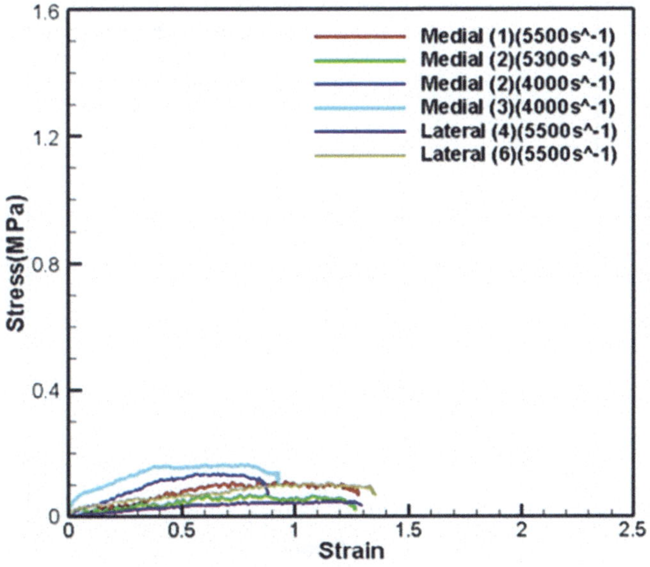

Fig. 2.5 Dynamic shear meniscus circumferentially cut the stress-strain curve

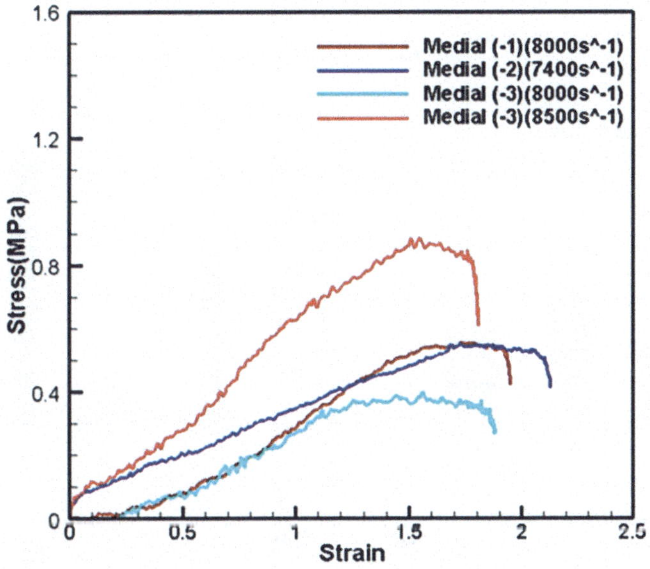

Fig. 2.6 Dynamic shear stress curve of medial cartilage

3. The maximum shear strength and the shear modulus of the medial side tend to be higher than that of the lateral meniscus, most notably in the axial incision. This result shows why most of the injuries to the medial meniscus in sports injuries are more likely than the outside.

Acknowledgement This research was supported by the Taiwan National Science Council under grant no. 105-2221-E-151-015, and National Kaohsiung University of Science and Technology, Taiwan.

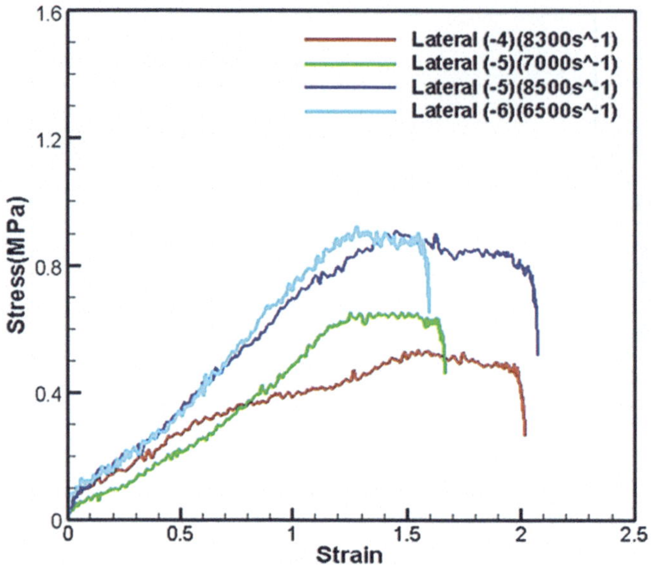

Fig. 2.7 Dynamic shear lateral cartilage stress-strain curve

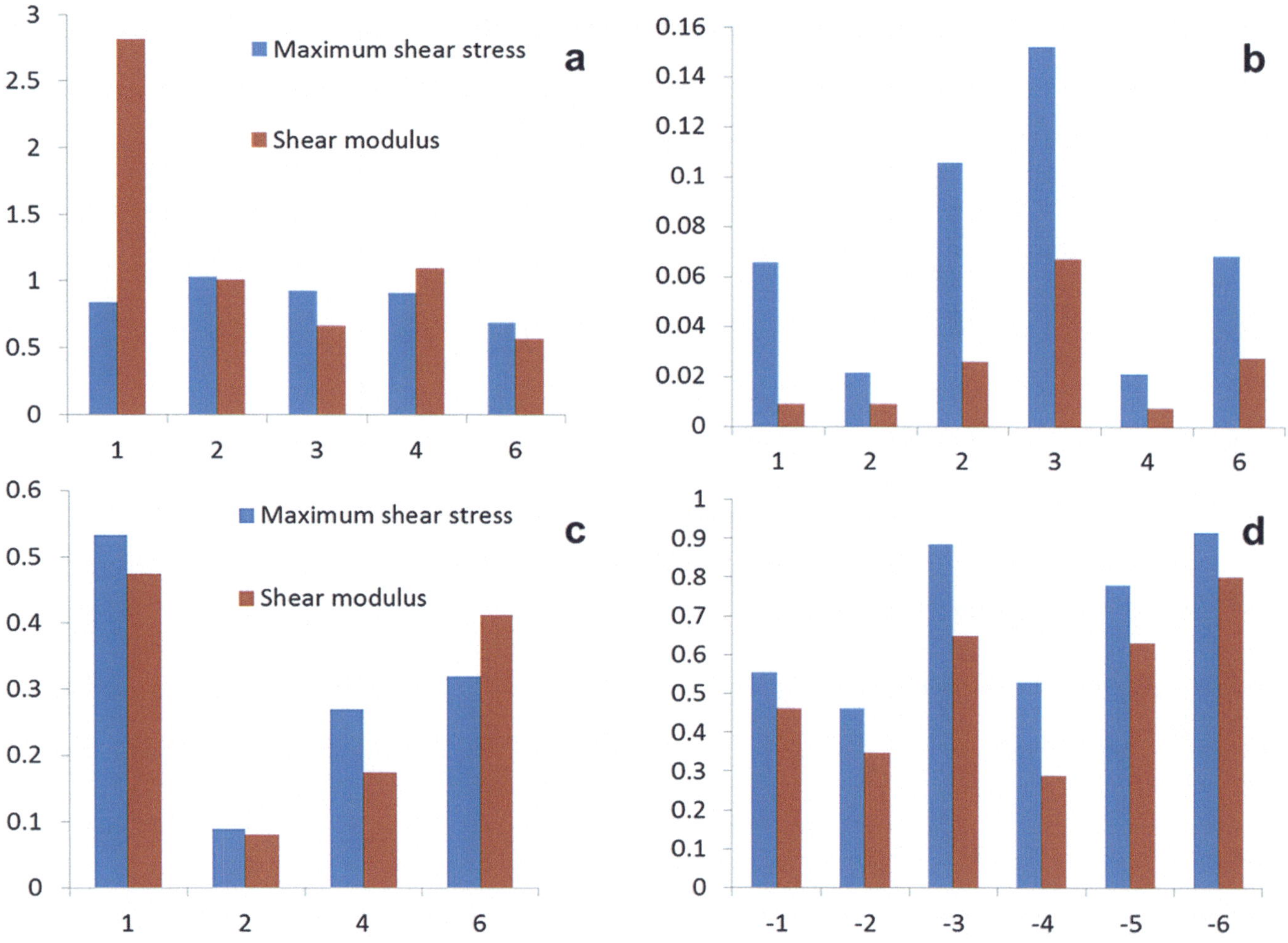

Fig. 2.8 Different types of cutting method of maximum shear stress and modulus comparison chart. (**a**) Meniscus axial cut; (**b**) radius meniscus cut; (**c**) meniscus circumferentially cut; (**d**) cartilage in all parts

References

1. Shephard, R.J.: Aging, Physical Activity, and Health. Human Kinetics Publishers, Champaign (1997)
2. Felson, D.T.: Epidemiology of hip and knee osteoarthritis. Epidemiol. Rev. **10**, 1–28 (1988)
3. Proffen, B.L., McElfresh, M., Fleming, B.C., Murray, M.M.: A comparative anatomical study of the human knee and six animal species. Knee. **19**, 493–499 (2012)
4. Zhu, W., Chern, K.Y., Mow, V.C.: Anisotropic viscoelastic shear properties of bovine meniscus. Clin. Orthop. Relat. Res. **306**, 34–45 (1994)
5. Messner, K., Gao, J.Z.: The meniscus of the knee joint. Anatomical and functional characteristics, and a rationale for clinical treatment. J. Anat. **193**, 161–178 (1998)
6. Ahmed, A.M., Burke, D.L.: In-vitro measurement of static pressure distribution in synovial joints—part 1: tibial surface of the knee. J. Biomech. Eng. **105**(3), 216–225 (1983)
7. Chevrier, A., et al.: Meniscus structure in human, sheep, and rabbit for animal models. J. Orthop. Res. **27**(9), 1197–1203 (2009)
8. Ratzlaff, C.R., Liang, M.H.: New developments in osteoarthritis. Prevention of injury-related knee osteoarthritis: opportunities for the primary and secondary prevention of knee osteoarthritis. Arthritis Res. Ther. **12**(4), 215 (2010)
9. Lynn, S.K., Reid, S.M., Costigan, P.A.: The influence of gait pattern on signs of knee osteoarthritis in older adults over a 5-11 year follow-up period: a case study analysis. Knee. **14**, 22–28 (2007)
10. Das, P., Schurman, D.J., Smith, R.L.: Nitric oxide and G proteins mediate the response of bovine articular chondrocytes to fluid-induced shear. J. Orthop. Res. **15**, 87–93 (1997)
11. Lee, M.S., Trindade, M.C.D., Ikenoue, T., et al.: Effects of shear stress on nitric oxide and matrix protein gene expression in human osteoarthritic chondrocytes in vitro. J. Orthop. Res. **20**, 556–561 (2002)
12. Smith, R.L., Donlon, B.S., Gupta, M.K., et al.: Effects of fluid-induced shear on articular chondrocyte morphology and metabolism in vitro. J. Orthop. Res. **13**, 824–831 (1995)
13. Tomatsu, T., Imai, N., Takeuchi, N., et al.: Experimentally produced fractures of articular-cartilage and bone—the effects of shear forces on the porcine knee. J. Bone Joint Surg. Br. **74**, 457–462 (1992)
14. Anderson, D.R., et al.: Viscoelastic shear properties of the equine medial meniscus. J. Orthop. Res. **9**(4), 550–558 (1991)
15. McDevitt, C.A., Webber, R.J.: The ultrastructure and biochemistry of meniscal cartilage. Clin. Orthop. Relat. Res. **282**, 8–18 (1990)
16. Rattner, J.B., et al.: New understanding of the complex structure of knee menisci: implications for injury risk and repair potential for athletes. Scand. J. Med. Sci. Sports. **21**(4), 543–553 (2011)

L. Tsai Associate Professor of National Kaohsiung University of Science and Technology, Kaohsiung with expertise in Materials Engineering, Mechanical Engineering, Structural Engineering.

Chapter 3
Tensile Response of Ceramics at the Microscale

Daniel J. Magagnosc and Brian E. Schuster

Abstract The tensile strength of ceramic materials are well known to be highly size dependent, arising from the sampling of intrinsic flaws. As such, determining the true tensile strength of ceramics remains a challenge. Here, we investigate the distribution of tensile properties in armor ceramics (SiC, B_4C, and SiC-B_4C composite) at the microscale. By employing high throughput femtosecond laser based machining, tensile bars with critical dimensions less than 100 μm are fabricated. The specimens are tested in a custom micromechanical testing apparatus in order to create a statistically significant failure distribution. The fracture surfaces are further characterized to assess the failure mechanisms. Together these observations provide a clear picture of the intrinsic tensile response of armor ceramics.

Keywords Fracture strength · Micro-machining · Micro-mechanics · Silicon carbide · Size effects

High strength ceramics are particularly susceptible to rupture upon the development of tensile stresses. In such cases, failure is understood to originate from activation of the "weakest link" flaw. As a consequence, ceramics typically do not exhibit a single tensile strength. Rather, a distribution of strengths is measured, which typically show a strong size dependence [1, 2]. A detailed examination of the tensile strength distribution is particularly pertinent for armor ceramics [3, 4]. Here, the tensile properties of SiC-N are investigated at the microscale in order to probe the intrinsic response of the material.

The micromechanical of investigation of the SiC-N is enabled through the use of a custom femtosecond (fs) laser milling system [5, 6]. Commercially sourced SiC-N is first cut and lapped to 13 mm × 13 mm × 0.2 mm thick coupons. The sections are then polished to a final thickness of 50 μm and a RMS surface roughness of <20 nm. The fs-laser milling system is subsequently used to cut tensile specimens from the thin sections. Here, a gently curved gage section, adapted from Sharpe et al. [7], is employed to minimize stress concentrations at the grip. After machining, the specimens are tested in tension using a custom micromechanical testing apparatus, which consists of translation stages for proper sample alignment, an S-beam load cell, and an optical microscope. Strains are applied using a precision linear motor stage while simultaneously measuring the force and capturing optical micrographs. Finally, the fractures surfaces are examined by scanning electron microscope.

In total, 38 tensile tests were performed on SiC-N. An optical micrograph of a representative tensile bar is shown in Fig. 3.1a. Here, the total gage length is 700 μm and the minimum gage width is ~50 μm. The resulting specimen volume is ~0.003 mm^3. The corresponding stress-strain curves and fracture surface are shown in Fig. 3.1b. The individual loading curves are offset for clarity. The stress-strain curves exhibit the expected brittle tensile response. Similarly, the fracture surfaces show clear evidence of brittle failure, primarily via intergranular fracture. Finally, a Weibull plot for the fracture strengths is shown in Fig. 3.1c. The resulting fit indicates a Weibull modulus of 6.63 and a characteristic strength of 615 MPa; the average strength is 574 MPa. When compared with previous reports of SiC strength [3, 4], the microtensile results suggest a saturation in fracture strength below a critical length scale on the order of 1 mm^3. The saturation likely results from an exhaustion of macroscopic flaws, which dominate the response at larger length scales.

D. J. Magagnosc (✉) · B. E. Schuster
US Army Research Laboratory, Aberdeen Proving Ground, MD, USA
e-mail: daniel.j.magagnosc.ctr@mail.mil

M. Grady et al. (eds.), *Mechanics of Biological Systems & Micro-and Nanomechanics, Volume 4*, Conference Proceedings of the Society for Experimental Mechanics Series, https://doi.org/10.1007/978-3-319-95062-4_3

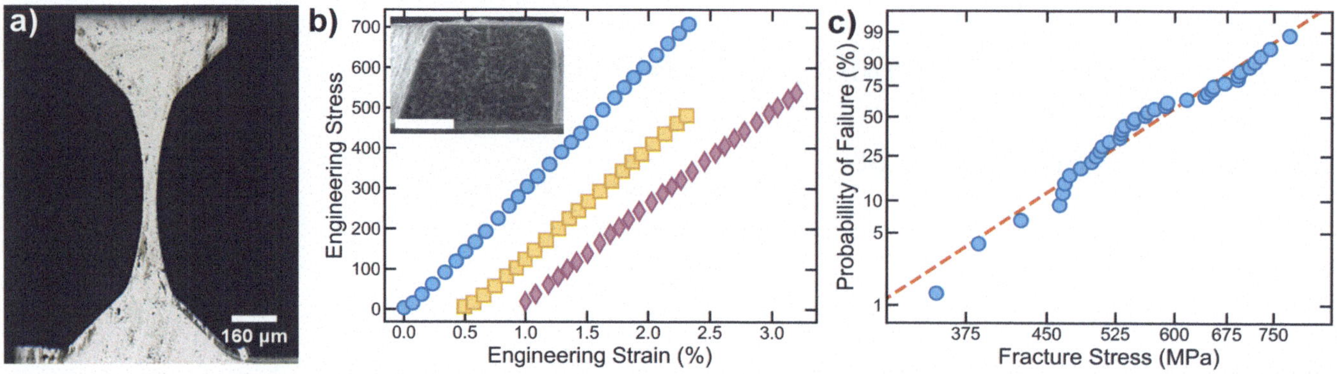

Fig. 3.1 Microscale tensile tests are performed on fs-laser machined specimens. A gently curved gage section is employed to minimize unwanted stress concentrations in the grip region. An example test specimen is shown in (**a**). Representative engineering stress-strain curves indicates brittle failure and are presented in (**b**). Here, the curves are offset for clarity. The inset shows a representative fracture surface (scale bar is 30 μm), which indicates predominantly intergranular fracture. Finally, a Weibull plot of the measure fracture strengths is determined in (**c**). The resulting Weibull modulus and characteristic strength are 6.63 and 615 MPa, respectively

References

1. Doremus, R.H.: Fracture statistics: a comparison of the normal, Weibull, and type I extreme value distributions. J. Appl. Phys. **54**, 193–198 (1983)
2. Lu, C., Danzer, R., Fischer, F.D.: Fracture statistics of brittle materials: Weibull or normal distribution. Phys. Rev. E Stat. Nonlinear Soft Matter Phys. **65**, 067102 (2002)
3. Strack, O.E., Leavy, R.B., Brannon, R.M.: Aleatory uncertainty and scale effects in computational damage models for failure and fragmentation. Int. J. Numer. Methods Eng. **102**, 468–495 (2015)
4. Wereszczak, A.A., Kirkland, T.P., Strong, K.T., Campbell, J., LaSalvia, J.C., Miller, H.T.: Size-scaling of tensile failure stress in a hot-pressed silicon carbide. Int. J. Appl. Ceram. Technol. **7**, 635–642 (2010)
5. Magagnosc, D.J., Ligda, J.P., Sano, T., Schuster, B.E.: Femtosecond laser machining of micro-tensile specimens for high throughput mechanical testing. In: Starman, L., Hay, J. (eds.) Micro and Nanomechanics. Conference Proceedings of the Society for Experimental Mechanics Series, vol. 5, pp. 7–9. Springer, Cham (2018)
6. Slaughter, S.K., Ligda, J.P., Sano, T., Schuster, B.E.: High throughput femtosecond-laser machining of micro-tension specimens. In: TMS 2015, 144th TMS Annual Meeting & Exhibition, Supplemental Proceedings, Orlando, pp. 471–478. Wiley, Hoboken (2015)
7. Sharpe, W.N., Jadaan, O., Beheim, G.M., Quinn, G.D., Nemeth, N.N.: Fracture strength of silicon carbide microspecimens. J. Microelectromech. Syst. **14**, 903–913 (2005)

Daniel J. Magagnosc received his Ph.D. in Materials Science from the University of Pennsylvania. His research interests focus on using in situ testing methodologies to understand deformation mechanisms in microscale materials.

Chapter 4
Development of Femtosecond Laser Based Microscale Fracture Methods

Daniel J. Magagnosc and Brian E. Schuster

Abstract Growing interest in characterizing the fracture response in non-bulk systems, such as second phase precipitates or microscale structural materials, necessitates the development of accurate fracture methodologies at relevant length scales. Here, femtosecond laser based micromachining is employed to fabricate microscale fracture specimens. Femtosecond laser machining enables rapid production of fractures specimens with sharp initial notches. Fracture tests are subsequently performed in a custom micromechanical testing apparatus. Initially, the femtosecond laser based fracture methodology is applied to fused quartz to validate the accuracy of the measurements. Further measurements are made on polycrystalline silicon carbide to demonstrate the broader applicability to higher fracture toughness materials.

Keywords Fracture toughness · Micro-machining · Micro-mechanics · Glass · Silicon carbide

As interest in characterizing the fracture toughness of second phases and other materials at small length scale has grown, numerous fabrication and measurement methodologies have been developed. Many of these methods have been adapted from bulk techniques and can be broadly characterized as indention, tension, or micro-cantilever bending, with cantilever bending being most prevalent [1–5]. Yet, many of the techniques suffer from poor scalability and low throughput, stemming from the low material removal rate inherent to focused ion beam (FIB) based fabrication. In order to accelerate the fabrication and testing process a new femtosecond laser-based microscale fracture testing method is suggested. Here, this process is demonstrated on fused quartz and silicon carbide.

Microscale fracture specimens are fabricated using a custom fs-laser machining system. Commercially sourced fused quartz and silicon carbide (SiC-N), is lapped and polished to a final thickness of 50 μm and a RMS surface roughness of less than 2 and 20 nm, respectively. Using the fs-laser, double edge notch tension (DENT) test specimens are machined using an average continuous power of 10 mW. The target gage width (W) is 150 μm with an initial crack length of 0.55W. Furthermore, a long gage length is employed to minimize confining stresses during testing. After laser machining, the samples are transferred to a custom micromechanical testing apparatus, which consists of an S-beam load cell, optical microscope, and micropositioning stages for sample alignment. The samples are subsequently loaded using precision motor stages; the load on the sample and optical micrographs are concurrently captured using custom Labview scripts.

An optical micrograph of a representative DENT machined in fused quartz is shown in Fig. 4.1a. In total, two successful fracture tests were performed on each material. The observed fracture toughness, K_{IC}, for each specimen is shown in Fig. 4.1b. K_{IC} was calculated from the maximum force and the measured, gage thickness, gage width, and initial crack length according to Tada et al. Overall, the measure K_{IC} from fused quartz (0.81 and 0.77 MPa-m$^{0.5}$) shows excellent agreement with the bulk value of 0.787 MPa-m$^{0.5}$ [6]. In contrast, the measured K_{IC} for SiC-N (1.83 and 3.83 MPa-m$^{0.5}$) shows significant variation. However, the maximum observed K_{IC} of 3.83 MPa-m$^{0.5}$ approaches the expected bulk value of 4–5 MPa-m$^{0.5}$ [7]. Taken together, these observation from fused quartz and SiC-N suggest that fs-laser based microscale fracture toughness methods are a promising experimental avenue.

D. J. Magagnosc (✉) · B. E. Schuster
US Army Research Laboratory, Aberdeen Proving Ground, MD, USA
e-mail: daniel.j.magagnosc.ctr@mail.mil

M. Grady et al. (eds.), *Mechanics of Biological Systems & Micro-and Nanomechanics, Volume 4*, Conference Proceedings of the Society for Experimental Mechanics Series, https://doi.org/10.1007/978-3-319-95062-4_4

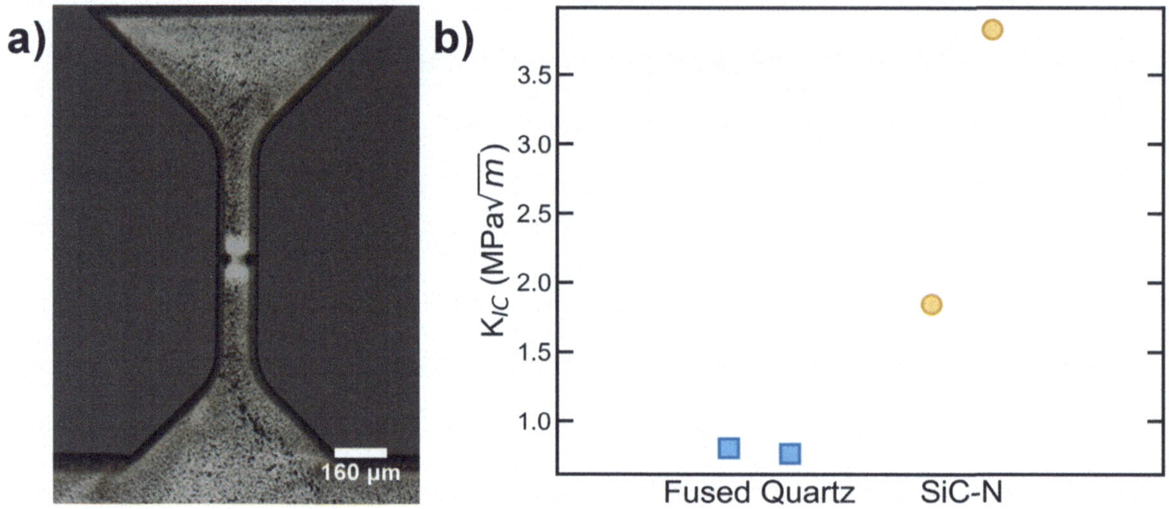

Fig. 4.1 (**a**) An example double edge notch tension specimen, produced by fs-laser micromachining in fused quartz. (**b**) Measured fracture toughness in fused quartz and SiC-N. The fused quartz shows excellent agreement with the expected bulk fracture toughness. However, while the measured SiC-N toughness approaches bulk values, significant variability is observed

References

1. Casellas, D., Caro, J., Molas, S., Prado, J.M., Valls, I.: Fracture toughness of carbides in tool steels evaluated by nanoindentation. Acta Mater. **55**, 4277–4286 (2007)
2. Espinosa, H.D.: A new methodology to investigate fracture toughness of freestanding MEMS and advanced materials in thin film form. J. Microelectromech. Syst. **14**, 153–159 (2005)
3. Jaya, B.N., Kirchlechner, C., Dehm, G.: Can microscale fracture tests provide reliable fracture toughness values? A case study in silicon. J. Mater. Res. **30**, 686–698 (2015)
4. Mueller, M.G., Pejchal, V., Žagar, G., Singh, A., Cantoni, M., Mortensen, A.: Fracture toughness testing of nanocrystalline alumina and fused quartz using chevron-notched microbeams. Acta Mater. **86**, 385–395 (2015)
5. Žagar, G., Pejchal, V., Mueller, M.G., Michelet, L., Mortensen, A.: Fracture toughness measurement in fused quartz using triangular chevron-notched micro-cantilevers. Scr. Mater. **112**, 132–135 (2016)
6. Lucas, J.P., Moody, N.R., Robinson, S.L., Hanrock, J., Hwang, R.Q.: Determining fracture toughness of vitreous silica glass. Scr. Metall. Mater. **32**, 743–748 (1995)
7. Wereszczak, A.A., Johanns, K.E., Jadaan, O.M.: Hertzian ring crack initiation in hot-pressed silicon carbides. J. Am. Ceram. Soc. **92**, 1788–1795 (2009)

Chapter 5
Programming Vanadium Dioxide Based MEMS Mirror

D. Torres, Sarah Dooley, La Vern Starman, and Nelson Sepúlveda

Abstract The programming of different mechanical states of the tip-tilt piston vanadium dioxide (VO_2)-based MEMS micromirror is presented. VO_2 is a smart material that exhibits a solid-to-solid phase change transition at around 68 °C which spans about 10 °C. During the transition the mechanical properties of the material change, inducing a stress on the microstructure generating a displacement. The micromirror device has a fill-factor of 32.8% and is actuated through Joule heating, using monolithically integrated resistive heaters. The hysteretic behavior of the VO_2 is exploited to program either rotational angles (tilt-mode) or vertical displacements (piston-mode) of the device. The tilt-mode is divided in two components: pitch and roll. The programming of each mode is realized by applying a constant current to pre-heat the VO_2 film within the transition temperature and adding electrical pulses to move between states. The maximum power consumed when programming the two states is measured for each actuation mode. The repeatability of the programming actions is demonstrated by showing an identical sequence of programming pulses produced the same mechanical state for the rotational angle (tilt-mode) and vertical displacement (piston-mode). Finally, it is demonstrated that different states can be programmed by using the same pre-heating temperature and programming pulses at different magnitudes.

Keywords MEMS · Micromirror · Vanadium dioxide · Phase-change materials · Programmable MEMS

5.1 Introduction

Microelectromechanical (MEMS) micromirrors have been designed and fabricated using conventional micromachining techniques and common MEMS actuation mechanisms such as: electrostatic (ES) [1], piezoelectric (PE) [2], electromagnetic (EM) [3], and electrothermal (ET) [4]. Each actuation mechanism has advantages and disadvantages, but all of them usually require a high input value (such as a voltage or a current) or have a high consumption power. This increases the complexity of incorporating micromirror devices with electronic circuits to create a standalone system. For example the ES and PE actuation mechanism requires a large voltage to generate the actuation while the power consumption is relatively low, due to the low current consumption [1, 2]. On the other side of the spectrum, EM actuation requires large current values for full actuation, with reported values of 515.17 mA [3]. The ET actuation does not require large input values, but instead is a combination of both: input voltage and current. This in turn increases the electrical power needed for full actuation, with reported values of 41 mW with an input voltage of 4.5 V [2] generating a temperature around 350 °C. Conventional ET designs rely on Joule heating and the coefficient of thermal expansion (CTE) of the materials to generate an actuation. For example, in [4] the CTE difference between the aluminum (Al) and silicon dioxide (SiO_2) is used to generate a stress capable of moving the structure as the temperature of the actuator is increased. In [5], we presented the first smart material based MEMS micromirror based on an ET mechanism by incorporating a vanadium dioxide (VO_2) layer into the structure. The VO_2 reduced the relatively large temperature gradient requirements in traditional CTE-based ET mechanisms from 350 °C to less than 100 °C, resulting in a reduction in power consumption. The device consisted of a mirror platform surrounded by four actuator legs, where the average power consumed between the actuators during individual actuation was 6.5 mW and for the generation of the vertical displacement (piston-like) was 26.1 mW.

D. Torres (✉) · S. Dooley · L. V. Starman
Sensors Directorate, Air Force Research Laboratory, Wright-Patterson AFB, OH, USA
e-mail: david.torres-reyes.1@us.af.mil

N. Sepúlveda
Department of Electrical and Computer Engineering, Michigan State University, East Lansing, MI, USA

M. Grady et al. (eds.), *Mechanics of Biological Systems & Micro-and Nanomechanics, Volume 4*, Conference Proceedings of the Society for Experimental Mechanics Series, https://doi.org/10.1007/978-3-319-95062-4_5

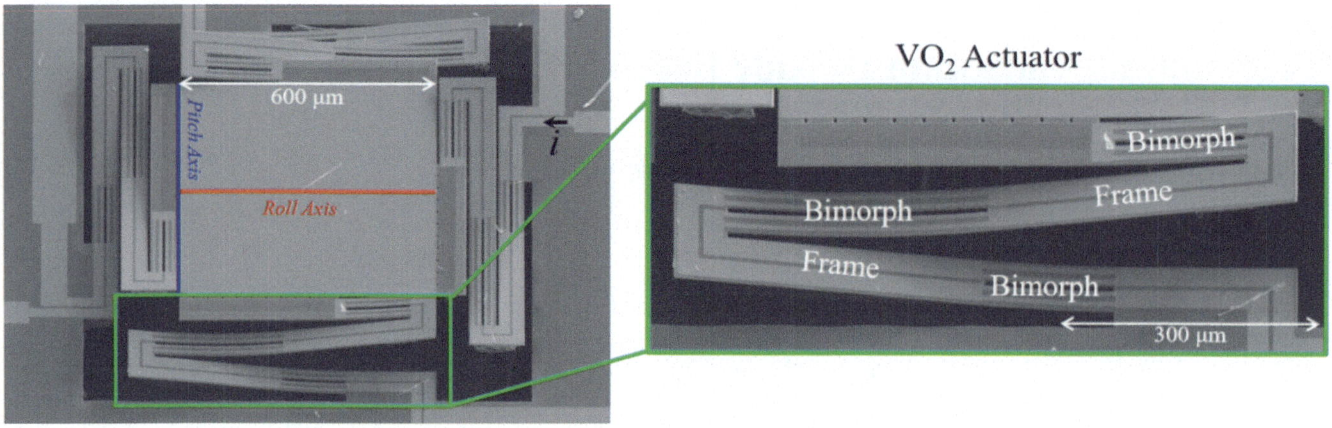

Fig. 5.1 SEM images of the VO_2-based MEMS mirror, where the rotational axes are labels for the actuated leg. The mirror platform is 600 μm by 600 μm, and the device size is 1.048 mm, resulting in a fill factor of 32.8%

VO_2 has a reversible solid-to-solid phase change transition with a transition temperature of around 68 °C and spanning about 10 °C. During the transition the mechanical properties of the material (such as the young's modulus and strain) drastically change [6, 7]. In [7], it was calculated that the strain energy density generated during the transition of the material was higher than the conventional actuation mechanisms implemented in MEMS, including: ES, PE, EM, ET. Furthermore, the material exhibits a hysteresis behavior between the mechanical properties and the temperature. Although this behavior is deemed undesirable due to the increase in complexity for control purposes, it can be used to create different mechanical states resulting in a programmable device.

5.2 Design and Operation of VO_2-Based MEMS Mirror

The design of the VO_2-based MEMS micromirror (shown in Fig. 5.1) is based on an electro-thermal tip-tilt piston MEMS mirror presented in [2], where the actuation mechanism (ET) is based on the difference in CTE of the materials in the bimorph regions. In the present work, the actuation mechanism can still be considered ET in that it uses an electro-thermal process. However, instead of relying on thermal expansion differences between the thin films composing the bimorph regions, mechanical actuation is mainly due to the thermally-induced solid-solid phase transition of the smart material coating (VO_2). Before and after the transition of VO_2, the dominant actuation mechanism is the difference of CTE between the layers, while stress levels induced during the phase-change of VO_2 dominates during the transition. The micromirror device used four individual actuators (legs) to move a square mirror platform. The legs are connected on each side of the micromirror platform, which results in a tilting movement upon the actuation of a single actuator; or the vertical platform displacement upon the simultaneous actuation of all four actuators.

The hysteresis behavior of the VO_2 is observed in Fig. 5.2 for tilt and piston mode. Although it is possible to program any state enclosed by the major hysteretic loop, for purposes of simplicity of the proof of concept, we focused the experiment on two different states: one above, and one below a pre-heated state. The experiment required three different current references: a constant current (i_{ph})—to remain inside the hysteresis; a high current value ($i+$)—to reach the high state; and a low current value ($i-$)—to reach the low state. It should be noted that a particular input programming sequence is applied for each mode.

5.3 Conclusion

A programmable VO_2 based MEMS micromirror is presented in this work. The device operation is based on the solid-to-solid phase transition of VO_2, and exhibits a lower actuation temperature than typical electrothermal MEMS mirrors based on thermal expansion coefficient mechanisms. Multiple mechanical states can be programmed in each of the two operating modes: tilt- and piston-mode. The tilt-mode consisted of two different angles: pitch and roll, while the piston-mode consisted of vertical displacements. Both modes are realized experimentally, demonstrating good repeatability and capability of programming multiple states.

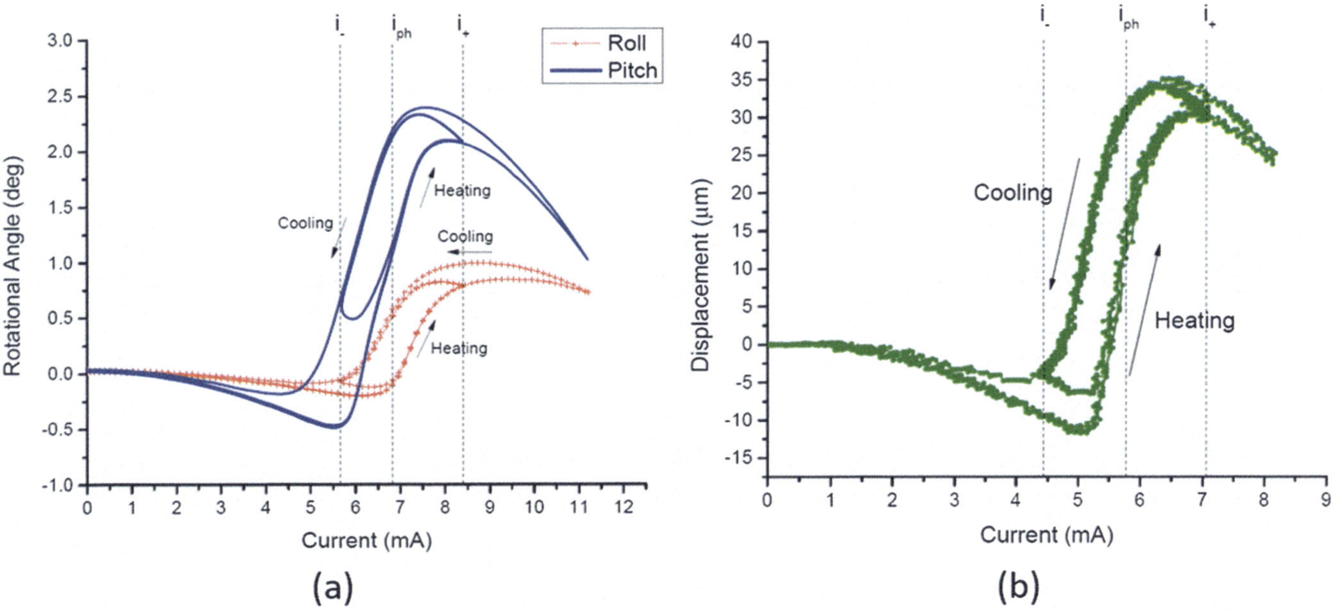

Fig. 5.2 Hysteretic behavior caused by the VO_2 with the respective input values used for the programming of two mechanical states, observed on both modes: (**a**) rotational angle of the tilt-mode where the input values are: $i_{ph} = 6.83$ mA, $i_- = 5.74$ mA and $i_+ = 8.39$ mA, (**b**) Vertical displacement of the piston-mode where the input values are: $i_{ph} = 5.78$ mA, $i_- = 4.42$ mA and $i_+ = 7.09$ mA. The minor loop in each plot represents the output history between the two states

Acknowledgment The fabrication of the MEMS Mirror was partially done at the Lurie Nanofabrication Facility at University of Michigan.

References

1. Hung, A.C.-L., Lai, H.Y.-H., Lin, T.-W., Fu, S.-G., Lu, M.S.-C.: An electrostatically driven 2D micro-scanning mirror with capacitive sensing for projection display. Sens. Actuators A Phys. **222**, 122–129 (2015)
2. Naono, T., Fujii, T., Esashi, M., Tanaka, S.: Non-resonant 2-D piezoelectric MEMS optical scanner actuated by Nb doped PZT thin film. Sens. Actuators A Phys. **233**, 147–157 (2015)
3. Cho, A.R., Han, A., Ju, S., Jeong, H., Park, J.-H., Kim, I., Bu, J.-U., Ji, C.-H.: Electromagnetic biaxial microscanner with mechanical amplification at resonance. Opt. Express. **23**(13), 16792–16802 (2015)
4. Wu, L., Dooley, S., Watson, E., McManamon, P.F., Xie, H.: A tip-tilt-piston micromirror array for optical phased array applications. J. Microelectromech. Syst. **19**(6), 1450–1461 (2010)
5. Torres, D., Wang, T., Zhang, J., Zhang, X., Dooley, S., Tan, X., Xie, H., Sepúlveda, N.: VO_2-based MEMS mirrors. J. Microelectromech. Syst. **25**(4), 780–787 (2016)
6. Sepúlveda, N., Rua, A., Cabrera, R., Fernández, F.E.: Young's modulus of VO_2 thin films as a function of temperature including insulator-to-metal transition regime. Appl. Phys. Lett. **92**, 191913 (2008)
7. Merced, E., Tan, X., Sepúlveda, N.: Strain energy density of VO_2-based microactuators. Sens. Actuators A Phys. **196**, 30–37 (2013)

D. Torres received a B.S. from the University of Puerto Rico at Mayagüez, and a Ph.D. from Michigan State University (MSU), both in electrical engineering. He is an electronics engineer in the Sensors Directorate at AFRL.

Chapter 6
Modeling and Simulation of Post Processed Foundry Fabricated Large, Out-of-Plane MEMS Energy Harvester

J. P. Walton, La Vern Starman, and D. Torres

Abstract This research effort is focused on the development of a micro-electro-mechanical system (MEMS) energy harvesting device using large, out-of-plane bimorph MEMS structures fabricated in the PolyMUMPs™ foundry process. In general, MEMS designers have limited options for commercial device fabrication, each bringing the advantage of quick prototyping, and repeatable results but also unique limitations. The design rules limit the designers to specific materials, structural layers, and fabrication processes to those of the fabrication facility. In previous work we have shown that the PolyMUMPs™ process will provide an adequate foundational structure for the device, meeting the initial out-of-plane deflection and flexibility requirements. However, it will require some post-processing to add the required mass and piezoelectric material layers to the structures. COMSOL™ finite element modeling software is used to evaluate various mass sizes, orientations, and materials as well as to explore different piezoelectric materials and configurations to produce optimal results from the in-house post-processing of these designs. Initial modeling of the structures illustrates unique design flexibility such that the operational frequency can be tailored to meet specific resonances enabling these devices to be adapted for a multitude of applications. Lastly, since the device provides the added benefit and flexibility, they can easily be integrated into an array for increased output voltage making them a prime candidate for micro power generation circuits.

Keywords MEMS · Energy harvesting · Bimorph · PolyMUMPs™ · Post processing

6.1 Introduction and Preliminary Results

With the growing interest in renewable energy micro-electro-mechanical system (MEMS) provide a feasible means of micro power generation for use in many consumer products. The challenge facing current MEMS designers revolves around the fabrication of designs which are optimized for power generation. Commercial fabrication foundries provide well-defined processes and repeatable results; however they generally have strict design constraints, sometimes limiting optimal material selection or structural considerations. Thus, this current research effort uses a commercially available PolyMUMPs™ foundry process to fabricate the basis structure of the energy harvesting device. Figures 6.1 and 6.2 illustrate the two design structures to be studied for energy harvesting MEMS devices. As both design approaches illustrate (see Figs. 6.1b and 6.2b), the PolyMUMPs™ foundry fabricated devices will require post processing through in-house techniques to increase the initial upward deflections as needed, to add the necessary mass to the platform, to deposit and integrate the piezoelectric material (PVDF, AlN, or PZT), and components into the device design. In order to complete the energy harvesting device the following material layers will be added in post-processing: an aluminum nitride piezoelectric layer of 1 μm, and a second gold layer with a thickness of 0.5 μm as the second electrode layer. This is followed by a 1 μm thick layer of Si_3N_4 exhibiting a stress level of 2.0 GPa, this Si_3N_4 layer results in an increase of the out-of-plane deflection of each device as it adds a bending moment in the critical locations, which ultimately helps increase the peak deflection. Figure 6.1 illustrates one example structure with anchors located at the center of the first beams. Figure 6.1b shows a scanning electron microscope (SEM) image of the fabricated, released structure, showing no out-of-plane deflection of the center platform. Figure 6.1c shows the modeled design including all the layers deposited (after the post-processing on the device). The out-of-plane deflection of the center platform was increased from ~12 to ~140 μm.

J. P. Walton · L. V. Starman · D. Torres (✉)
Sensors Directorate, Air Force Research Laboratory, Wright-Patterson AFB, OH, USA
e-mail: david.torres-reyes.1@us.af.mil

M. Grady et al. (eds.), *Mechanics of Biological Systems & Micro-and Nanomechanics, Volume 4*, Conference Proceedings of the Society for Experimental Mechanics Series, https://doi.org/10.1007/978-3-319-95062-4_6

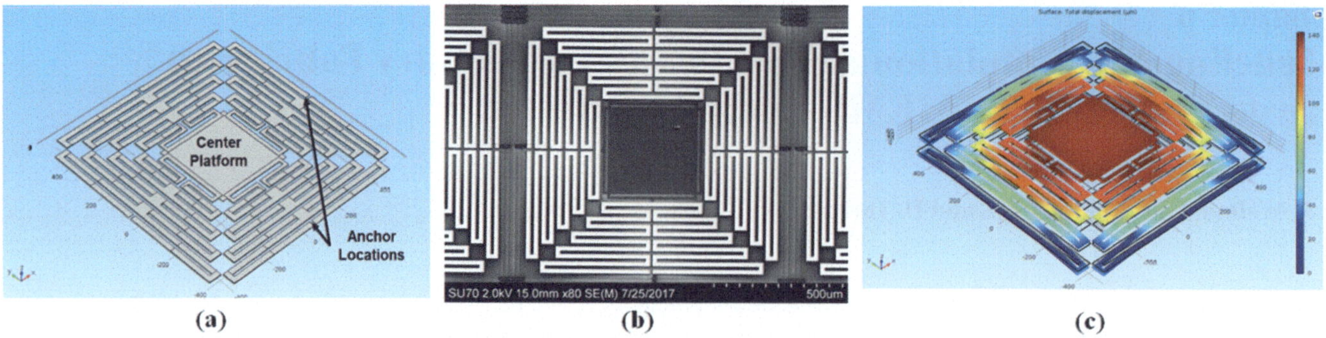

Fig. 6.1 (**a**) Representative energy harvesting MEMS structure where the systems anchors are at the center of each of the first beams as shown in the figure. (**b**) An SEM image of the released PolyMUMPs™ fabricated structure validating no out-of-plane deflection of the center platform. (**c**) A COMSOL™ image of the peak deflection of ~140 μm due to the post processing

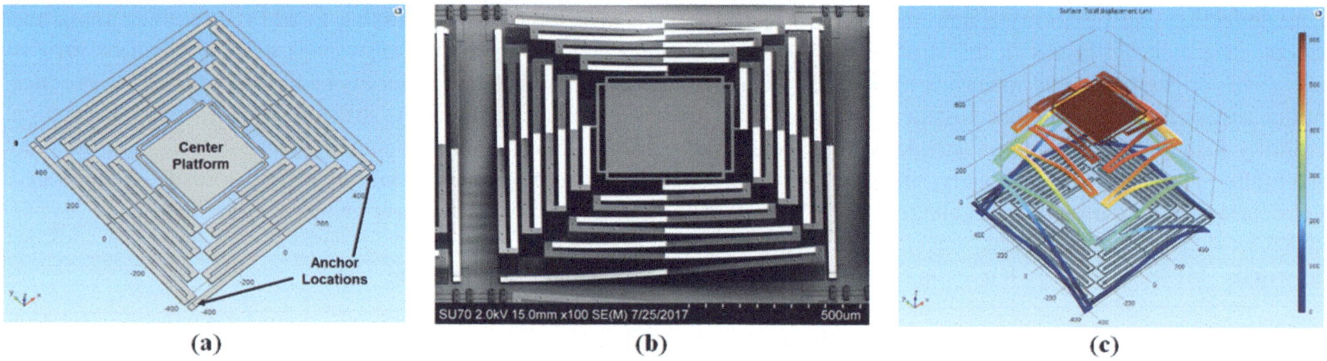

Fig. 6.2 (**a**) Image of the second energy harvesting design with the anchor positions noted in the image. (**b**) SEM image of a released PolyMUMPs™ of this design which exhibits ~148 μm of out-of-plane deflection. (**c**) A COMSOL™ representative image of the peak displacement for the proposed energy harvesting device with an out-of-plane deflection of ~600 μm

A second design is illustrated in Fig. 6.2 with the anchor positions shifted to the end of the first beam as shown in Fig. 6.2a. Figure 6.2b is an SEM image of the fabricated baseline structure in PolyMUMPs™ showing an out-of-plane deflection of the center platform as measured by white light interferometer (IFM) to be ~148 μm. Figure 6.2c shows the peak out-of-plane deflection for the post processing design in order to complete the energy harvesting device, with an out-of-plane deflection of over 600 μm.

Following the post processing of the Si_3N_4 layer at −2 GPa of stress on the design shown in Fig. 6.1, the initial out-of-plane deflection as measured with the IFM was ~170 μm as shown in Fig. 6.3a. This large deflection allows the freedom of movement required to maximize the harvesting of the kinetic energy while also enabling the ability to create arrays of these energy harvesting structures as shown in Fig. 6.3b to increase the overall generated output voltage. Figure 6.3b illustrates an array of post processed devices all with an upward deflection of ~170 ± 10 μm. This validates the process is repeatable and reliable for future post processing phases. The energy harvesting is optimal when the structure is operated in the piston mode (up and down); however, the tip/tilt mode will also produce energy but at a reduced level. Therefore, depending on the application, the mounting, orientation, and placement will need to be considered to maximize the energy harvesting.

The additional in-house processing allows for significant flexibility to tailor the device designs to pin-point specific resonant frequencies of interest by controlling the stress and displacement in the representative energy harvesting structures with defined, subsequent material layer placement and thicknesses [1–3]. The MEMS energy harvester design will need to maximize the out-of-plane initial upward deflection to maximize the piston motion of the structure while being constrained to a relatively small area (1 mm^2) [2]. The piston resonant frequency from COMSOL™ modeling results for the representative material layer stack as shown in Fig. 6.4a ranged from 28.7 kHz for the Fig. 6.1 design and 18.4 kHz for the Fig. 6.2 design. When a small mass ($3.712e^{-8}$ kg) was placed on the platform of both designs, the resonant frequencies reduced to 13.463 kHz for the Fig. 6.1 design and 6.939 kHz for the Fig. 6.2 design.

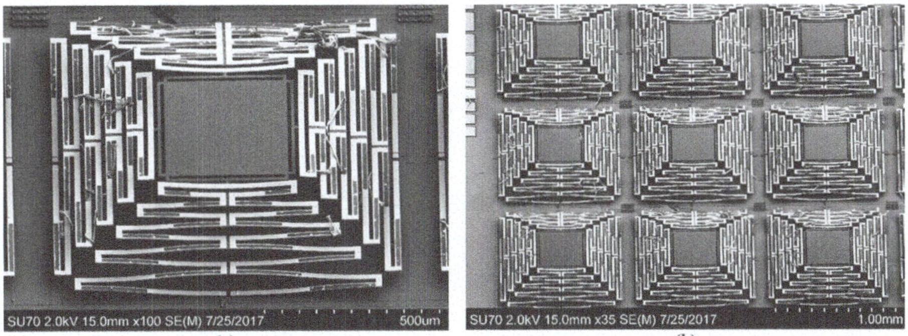

(a) (b)

Fig. 6.3 (**a**) A post processed SEM image of the PolyMUMPs™ design as shown in Fig. 6.1 with a 1 μm Si_3N_4 layer deposited on the structure to create the bending moment to enable an out-of-plane deflection of ~150 μm. (**b**) An SEM image of an array of candidate structures as shown in (**a**) which are post processed with Si_3N_4 showing a consistent upward out-of-plane deflection for integrating together to increase the overall voltage output of the energy harvesting system

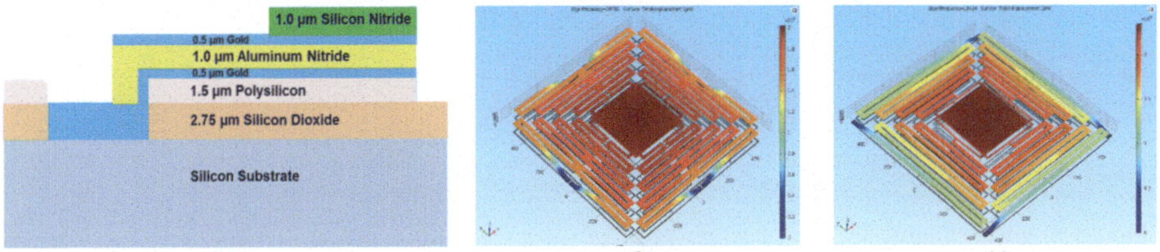

(a) (b) (c)

Fig. 6.4 (**a**) is a rendition of the material stack on the beam based structures, (**b**) shows the resonant frequency for the piston mode of the structure in Fig. 6.1 simulated to be 28.7 KHz, and (**c**) shows the resonant frequency for the piston mode of the structure in Fig. 6.2 simulated to be 18.4 KHz

6.2 Conclusions

The realization of MEMS based energy harvesting was presented which shows the PolyMUMPs™ fabricated structures can be configured through several post processing steps to provide a potential energy harvesting approach. As stated, these types of designs provide significant flexibility in tailoring the resonant frequency to the desired application. This tailoring can be accomplished in numerous ways to include changing materials, material thicknesses, number of beams, beam lengths, and the applied mass on the platform will all change the resonant frequency of the device. COMSOL™ models were used to evaluate these design structures and the post processing techniques. Future research will focus on the fabrication and test of these energy harvesting elements to determine output voltage levels as well as develop a representative in-house fabrication process which will allow for even greater design flexibility.

Acknowledgements Support for this research is greatly appreciated with funding through the Air Force Research Laboratory, Aerospace Component and Subsystem Division Titled: "Design, Model, and Fabrication of a Piston MEMS Micromirror Array".

References

1. Starman, L., Coutu Jr., R.: Stress monitoring of post-processed MEMS silicon microbridge structures using Raman spectroscopy. J. Exp. Mech. **52**(9), 1341–1353 (2011)
2. Walton, J.: Electrostatically driven large aperture micro-mirror actuator assemblies for high fill-factor, agile optical phase arrays. Defence Technical Information Center, ADA623146, 1–245, (2015)
3. Samuelson, S., Xie, H.: A large piston displacement MEMS mirror with electrothermal ladder actuator arrays for ultra-low tilt applications. J. Microelectromech. Syst. **23**(1), 39–49 (2014)

J. P. Walton received his M.S. degree in EE from AFIT in the focus area of MEMS and is currently pursuing his Ph.D. also from AFIT in EE. His current research focus areas include MEMS sensors and actuators, energy harvesting, and optical beamsteering.

Chapter 7
Measurement of the Viscoelastic Properties of the Chinchilla Tympanic Membrane

Junfeng Liang, Rong Z. Gan, and Hongbing Lu

Abstract Tympanic membrane (TM) transmits the sound pressure in the outer ear into mechanical vibration in the middle ear. Quantitate study on the middle ear sound transfer function requires accurate measurements of the TM mechanical properties. In this paper, a dynamic micro-fringe projection technique is developed to measure the viscoelastic properties of the chinchilla TM. An intact TM in the bulla is inflated with a time-varying air pressure. Micro-fringes are projected onto the TM surface during the inflation process and the surface image of the TM is captured with a digital video camera connected to a microscope. The pressure-volume displacement relationship is calculated with a phrase-shift algorithm. Finite element method with the implementation of a viscoelastic model is employed with the pressure history as input. A viscoelastic model is used to model material properties. By solving an inverse problem with the model, the viscoelastic properties of the TM are determined.

Keywords Tympanic membrane · Viscoelastic · Micro-fringe projection · Finite element method · Inverse method

7.1 Introduction

Studies of the mechanical behavior of the tympanic membrane (TM) are of a great interest since TM is the first and an important organ that transmits sound pressure to mechanical vibration into the inner ear. Recently, various high fidelity finite-element method (FEM) models of the middle ear have been developed to study the sound transmission and to simulate different types of auditory tests [1–3]. For these FEM models, the mechanical properties of the middle ear structures are the key impacts that control the accuracy of the simulation results. However, TM as a soft tissue has a highly complex laminate structure with nonlinear, anisotropic, time, and rate dependent behavior and is very sensitive to the testing condition. The mechanical properties of TM are difficult to measure and simulate. Many researchers have been actively developed techniques to measure the mechanical properties of TM. The existing measurement techniques, according to the sample preparation, generally fall into two categories: ex situ and in situ. In the former category, the TM was usually cut into strips or dissected from the middle ear and tested with various mechanical measurement settings such as tensile test [4], dynamic analysis [5], acoustic driving test [6], split Hopkinson tension bar [7], and nano-indentation [8]. With the help of imaging techniques, such as projection moiré [9], digital image correlation [4], and fringe projection [10, 11], moiré [12–15], in situ measurement of the mechanical properties of TM also became feasible by combing with indentation or pressure as loading.

In the most previous study, to simplify the measurement, TM was simplified as an elastic membrane. In recent years, viscoelastic and hyperelastic characteristics of TM were measured. However, so far there are no results reported on the mechanical properties of the TM with the consideration of both viscoelastic and hyperelastic characteristic. In this paper, we report our results on the visco-hyperelastic properties of TM using a dynamic micro-fringe projection technique. In this method, dynamic pressures were applied to the TM in situ and the images of the TM were acquired by a micro-fringe projection technique. FEM simulation results were used to solve an inverse problem to determine the visco-hyperelastic properties of the TM.

J. Liang · H. Lu (✉)
Department of Mechanical Engineering, University of Texas at Dallas, Richardson, TX, USA
e-mail: hongbing.lu@utdallas.edu

R. Z. Gan
School of Aerospace and Mechanical Engineering, University of Oklahoma, Norman, OK, USA

M. Grady et al. (eds.), *Mechanics of Biological Systems & Micro-and Nanomechanics, Volume 4*, Conference Proceedings of the Society for Experimental Mechanics Series, https://doi.org/10.1007/978-3-319-95062-4_7

7.2 Method

7.2.1 *Visco-Hyperelastic Model*

A large-strain visco-hyperelastic model was used to describe the mechanical behavior of the TM. This model is based on the formulation proposed by Simo et al. [16], amended to take into account the viscous shear response. The constitutive equations for hyperelastic material with strain-energy function Φ is given by:

$$S^{2d} = 2\frac{\partial \Phi}{\partial C} \tag{7.1}$$

where S^{2d} is the second Piola-Kirchhoff stress tensor and C is the Cauchy-Green deformation tensor. The true stress can be obtained as:

$$\sigma = \frac{1}{J} FS^{2d}F^T \tag{7.2}$$

Where F is the deformation gradient tensor and J is the determinant of F. Following Simo, et al. [12] and Holzapfel et al. [17], assuming the material is isotropic, the viscoelastic constitutive equations, in terms of the second Piola-Kirchhoff stress, is given by

$$S^{2d} = \int_0^t \left[\alpha_\infty^K + \sum_{i=1}^{n} \alpha_i^K \exp\left(-\frac{t-\tau}{\tau_i^K} \right) \right] \left(2\frac{d}{d\tau}\frac{dU}{dJ} \right) d\tau \tag{7.3}$$

where α_i^K are the bulk relative moduli and τ_i^K are the bulk relaxation time to be determined. U is the isotropic part of the strain energy. To simplify the parameter fitting, n was chosen as 1 in this study.

The constitutive model here chosen for the description of the hyperelastic behavior of the rubber is the second order Ogden model, as has been used in our previous work on the static measurement for guinea pig, chinchilla and human TMs. The Ogden formulation of the strain energy function U is expressed by:

$$U = \sum_{i=1}^{2} \frac{2\mu_i}{\alpha_i} \left(\lambda_1^{\alpha_i} + \lambda_2^{\alpha_i} + \lambda_3^{\alpha_i} \right), i = 1, 2 \tag{7.4}$$

where λ_j ($j = 1, 2, 3$) are the principle stretch ratios; μ_i and α_i are constants describing the hyperelastic properties of the material.

7.2.2 *Experimental Setup*

The schematics of experimental setup of the dynamic micro-fringe projection system is plotted in Fig. 7.1. The system consists of the loading section and the acquiring section. In the loading section, a syringe pump that consists of an actuator (Progressive Automation PA-15-411) and a pressure cylinder was used to apply pressure to the sample. In the acquiring section, a micro-fringe projector with a camera was used to capture the deformation of the sample and a pressure sensor (Omega Engineering, PX309) was used to record the history of the applied pressure. To synchronize the loading and the acquire system during the measurement, an electrically controlled shutter was also configured into the system. All the electrical driven components shared the same DC power supply (Topward, 6306D).

Before the test, the sample was connected to the syringe pump and the pressure sensor, and mounted on a sample holder that allowed a six degrees of freedom motion. The sample was positioned at the location with an angle so that the sample was covered with the fringes from the projector and could be acquired by the camera free from shadow. The camera was turned on to continuously record. The shutter was then manually set to close mode to mark the initial stage for image acquisition. When the power supply was turned on, the pressure sensor began to record the pressure. At the same time, the shutter open

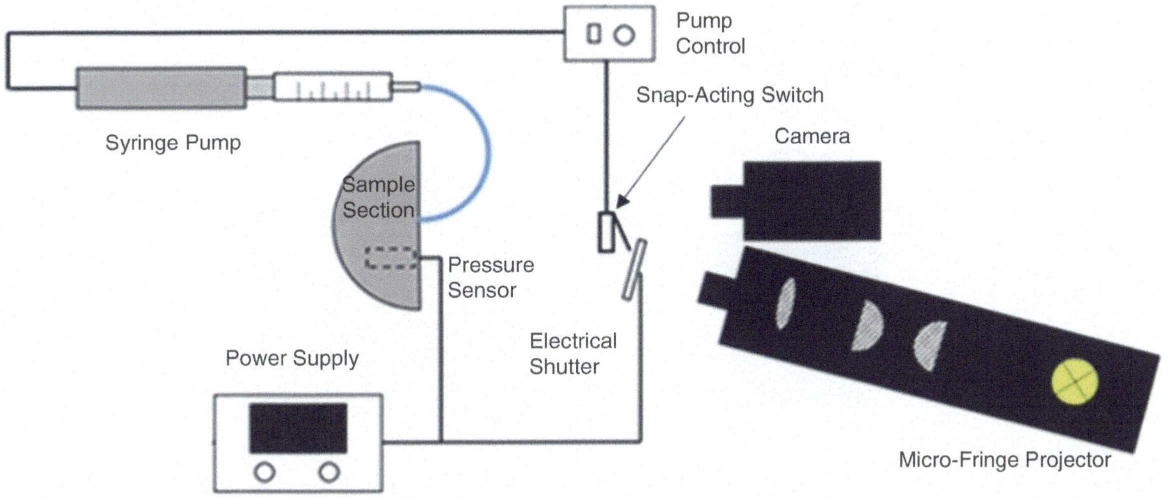

Fig. 7.1 Schematic of the Dynamic micro-fringe projection system. For illustration purpose, the components are not plotted with actual scale

to allow the camera to acquire the fringe pattern on the sample. It also simultaneously triggered a snap-acting switch that turned on the syringe pump. Therefore, the recorded data on pressure and the deformation were synchronized.

7.2.3 *Inverse Problem Solving*

The distorted micro-fringe patterns projected on a deformed surface contains the geometrical information of the surface. By comparing this pattern with that on a planar reference, with a five phase-shifted algorithm, the surface profile of the deformed surface of the sample under the pressure at each time step can be reconstructed. The volume displacement of the TM can be calculated by summation of the height over the surface as described with following formula:

$$\Delta V = \iint_{\Omega} [z(x, y) - z_0(x, y)]\, dxdy \tag{7.5}$$

where, $z(x,y)$ and $z_0(x,y)$ are the height profiles under a finite pressure and the zero pressure, respectively, Ω is the area enclosed by the annulus.

To determine the mechanical properties of the sample, an inverse problem solving procedure was used. The 3-dimensional surface reconstruction of the sample at zero-pressure state was used to build a corresponding FEM model. Simulation for the experiment was carried out by applying recorded pressure onto the model. Volume displacement from the simulation was compared with the experimental results to adjust the mechanical properties used in the simulation. This iteration continued until simulated volume displacement agreed well with the experimental one. The detail about the inverse problem can be found in our previous work [10, 11].

7.2.4 *Materials and Sample Preparation*

Latex rubber membrane (SensiCare, MD8085) was first used as a standard material to validate the method. Since there is no existed mechanical properties reported on the latex. The membrane material was measured with a dynamic tensile test to evaluate the viscoelastic properties. The data were then used to compare with that measured with the present dynamic micro-fringe projection measurement.

Figure 7.2 shows the configuration that used to hold the latex rubber membrane. The membrane was secured in a two part holder with a hole with 12.5 mm diameter that clamps the boundary of the membrane with a rubber o-ring. Such an arrangement allows the full sealant between the membrane and the hole in the holder and protects the membrane from being

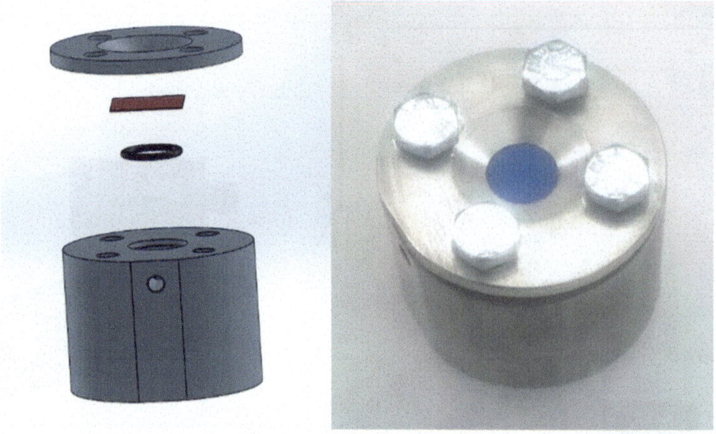

Fig. 7.2 Sample holder for testing rubber membrane using a dynamic micro-fringe projection system

Table 7.1 Dimensions of the chinchilla TM specimens

Sample #	Ear	Superior-inferior (mm)	Anterior-posterior (mm)
Ch17-2-6	R	8.16	9.15
Ch17-2-8L	L	7.89	9.32
Ch17-2-8R	R	8.06	8.18
Ch17-2-11	R	7.64	8.89
Ch17-2-13	L	8.08	8.67
Ch17-4-2L	L	8.13	9.06
Ch17-4-3	L	8.07	8.90
Ch17-4-4	L	7.99	8.91
Ch17-4-13	L	8.02	9.03

damaged by the shearing force from the holder. Pressure was applied to the hole of the holder through an outlet that connected to the syringe pump.

After validation of the method. Nine chinchilla TMs inside the bulla were tested using the dynamic micro-fringe projection system. The information of the samples was given in Table 7.1. The nine chinchilla bullas were harvest freshly from five adult chinchillas following the protocol approved by the Institutional Animal Care and Use Committee (IACUC) of the University of Oklahoma and met the guidelines of the National Institutes of Health. The TM was exposed for the projection and painted with a thin layer of titanium oxide to improve the reflectiveness. A polyvinyl chloride tubing was inserted and sealed on the ear canal. The TM and the ear canal formed a conceal chamber allowing the TM be inflated by the air pressure from the syringe pump. The total sample preparation time was about 45 min. A droplet of saline water was applied on the TM every 5 min to prevent the TM from dehydration. Before test, the sample was deflated and deflated several times at low pressure to allow the sample to reach steady state.

7.3 Result

7.3.1 Latex Rubber

Figure 7.3 shows the typical pictures of the latex rubber acquire by the camera. These pictures were extracted from a video on the deformation of the sample with ~3 s. To illustrate the deformation process, latex rubber under four different pressures at four different instances are plotted, which corresponding to the pressure history plot shown in Fig. 7.4.

Three-dimensional surface reconstruction was then applied on the images of the fringe patterns on the deformed membrane. Figure 7.5 demonstrates the surface profiles corresponding to the images in Fig. 7.3. Following Eq. (7.5), volume displacements over time were calculated with these surface profiles, which is shown in Fig. 7.6. Note that due the intensity of the data, only the first 0.6 s was used to determine the mechanical properties. Therefore, only ten points of volume

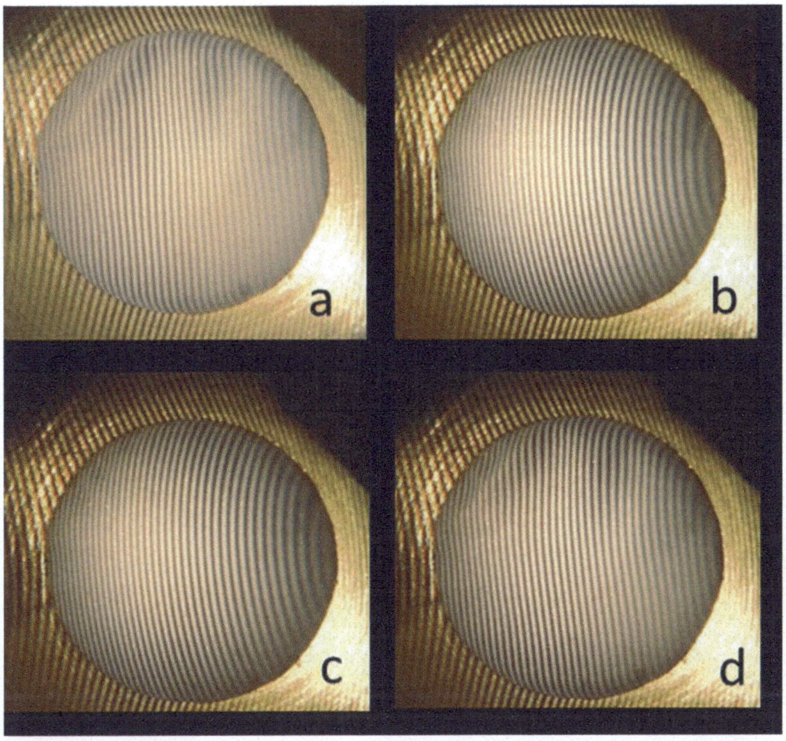

Fig. 7.3 Typical images of deformed latex rubber under air pressure at different time: (**a**) t = 0 s, p = 0 kPa; (**b**) t = 1 s, p = 2.3 kPa; (**c**) t = 2 s, p = 2.8 kPa; (**d**) t = 3 s, p = 0.7 kPa

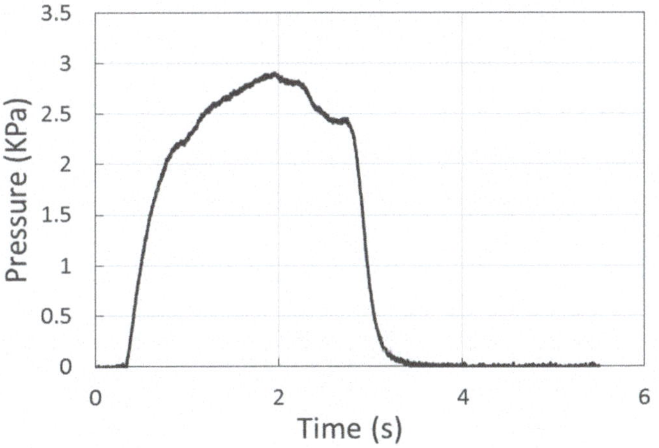

Fig. 7.4 History of pressure applied on the latex rubber

displacement were shown in the figure. FEM simulation of a membrane with the same dimension was run to determine the mechanical properties of the latex rubber using the iteration mentioned in the method section. The simulation results are also plotted in Fig. 7.6. The parameters used in the FEM simulation were: $\mu_1 = 0.4$, $\alpha_1 = 3.0$, $\mu_2 = 4.3$, $\alpha_2 = -0.9$, $\gamma = 0.15$, and $\tau = 950$. To validate the mechanical properties measured, the results are compared with that measured with a tensile test. Figure 7.7 shows the comparison between results from micro-fringe projection measurement and the tensile test on the latex rubber membrane.

Following the same procedure, dynamic micro-fringe measurements were applied on the chinchilla TM specimens. The history of the pressures applied on the nine chinchilla TM specimens were plotted in Fig. 7.8. The magnitude of the pressures and the time span were currently randomly controlled manually. The actually pressure histories were used in the FEM simulation. Figure 7.9 shows the specimen under different pressures projected with micro-fringes and their corresponding 3-dimensional surface reconstruction. Using the surface profile, the volume displacements of the nine chinchilla TMs were

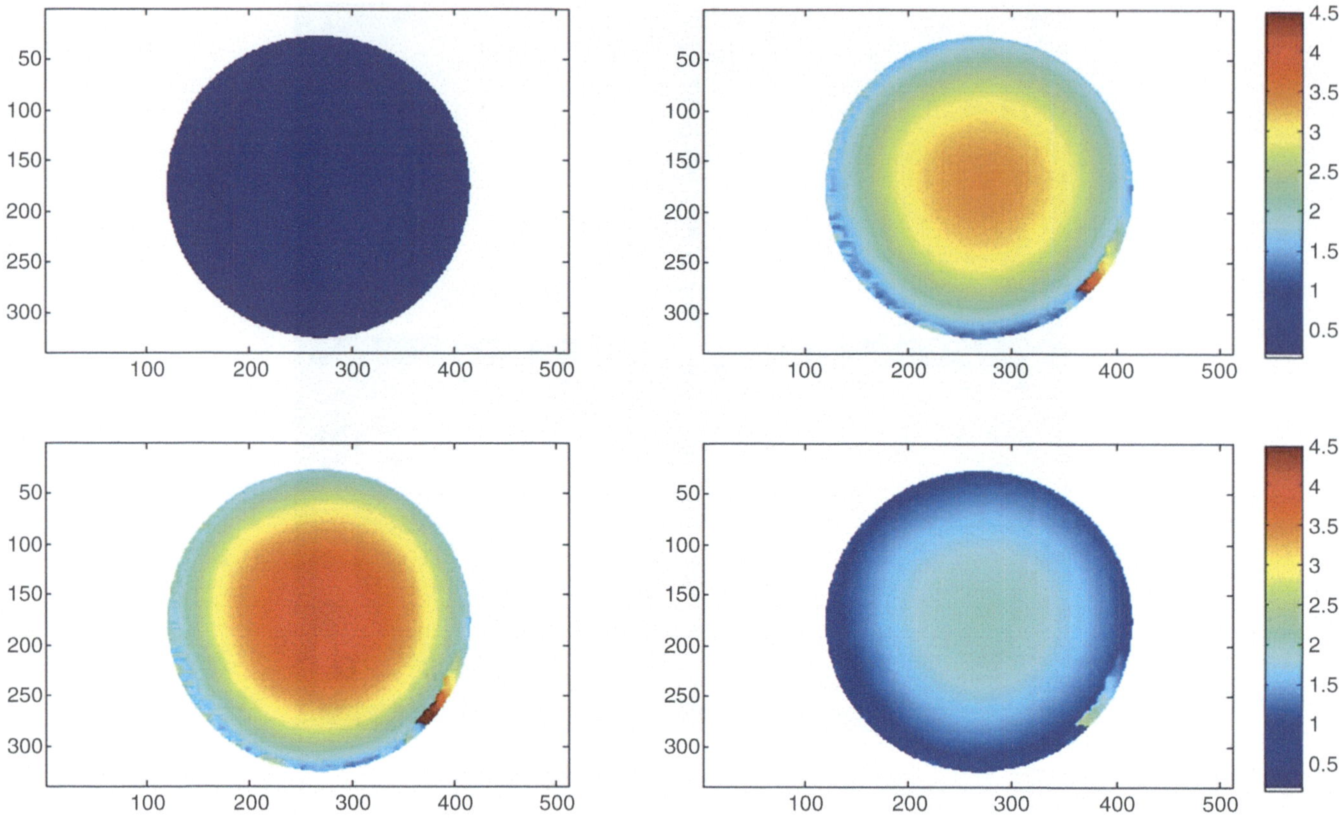

Fig. 7.5 Three-dimensional surface tomography of the deformed latex rubber: (**a**) t = 0 s, p = 0 kPa; (**b**) t = 1 s, p = 2.3 kPa; (**c**) t = 2 s, p = 2.8 kPa; (**d**) t = 3 s, p = 0.7 kPa. All the numbers in the figure are all in the unit of millimeter

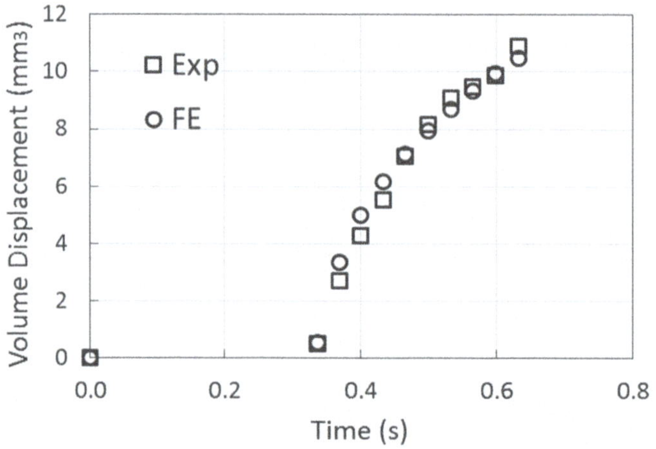

Fig. 7.6 Curve-fitting for the volume displacement of the latex rubber versus time: Exp (Experimental results) and FE (Finite element results)

calculated and were shown in Fig. 7.10. Note that volume displacements of at only ten different pressures were shown in the figure. FEM simulation was run for the inverse problem on each specimen to determine the mechanical properties using the constitutive law in Eq. (7.3). The curve fitting in the inverse problem were plotted in Fig. 7.10 in dash lines. The mechanical determined properties were listed in Table 7.2. The first four parameters describe the hyperelastic behavior of the TM materials as shown in Eq. (7.4). The last two parameters describe the viscoelastic behavior of the TM, where γ is the relative modulus and τ is the relaxation time.

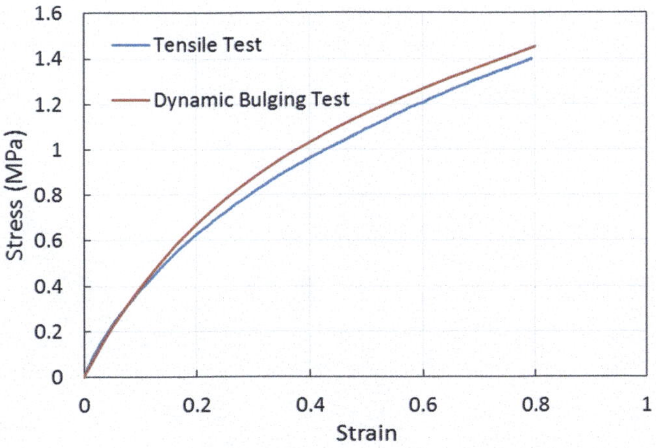

Fig. 7.7 Mechanical properties of latex rubber measurement by tensile test and dynamic bulging test micro-fringe projection system

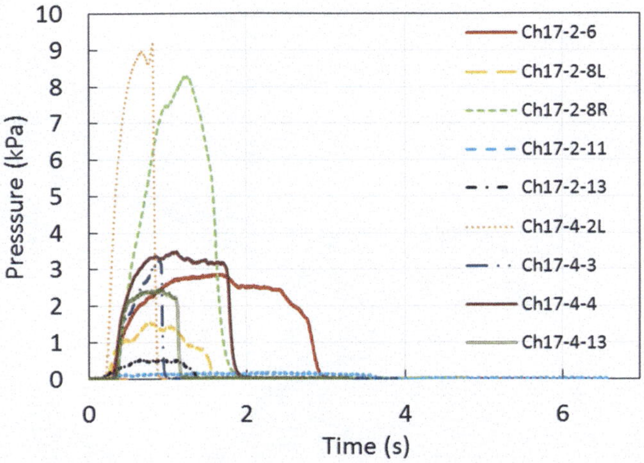

Fig. 7.8 The histories of the pressures applied on the nine chinchilla TM specimens

Table 7.2 Visco-hyperelastic properties of chinchilla TMs

Sample #	μ_1	α_1 (MPa)	μ_2	α_2 (MPa)	γ	τ (s)
Ch17-2-6	2.7	7.0	1.8	−3.6	0.85	140
Ch17-2-8L	3.0	9.2	1.6	−5.8	1.02	220
Ch17-2-8R	2.4	8.1	2.1	−2.2	1.05	170
Ch17-2-11	1.4	6.1	1.9	−4.4	0.64	220
Ch17-2-13	3.7	4.8	1.6	−4.7	0.89	190
Ch17-4-2L	3.4	7.8	1.9	−5.8	0.89	260
Ch17-4-3	3.2	6.6	2.1	−2.7	1.07	210
Ch17-4-4	2.8	5.4	2.1	−3.5	1.04	90
Ch17-4-13	2.8	8.4	2.1	−4.1	0.96	190

7.4 Discussion

In this paper, we present a new method, dynamic micro-fringe projection, to simultaneously determine the viscoelastic and hyperelastic properties of TM. Latex Rubber was used as a standard material to validate the setup. In Fig. 7.7 it's shown that the mechanical properties determined by the dynamic micro-fringe were slightly higher than that measured from a standard tensile test. At the maximum strain 0.8, the stress determined by dynamic micro-fringe was approximately 7% higher than that measured from tensile test. The difference of the mechanical responses measured from the two methods could stem from the difference of the strain rate: in dynamic micro-fringe projection measurement. The loading in the dynamic micro-fringe

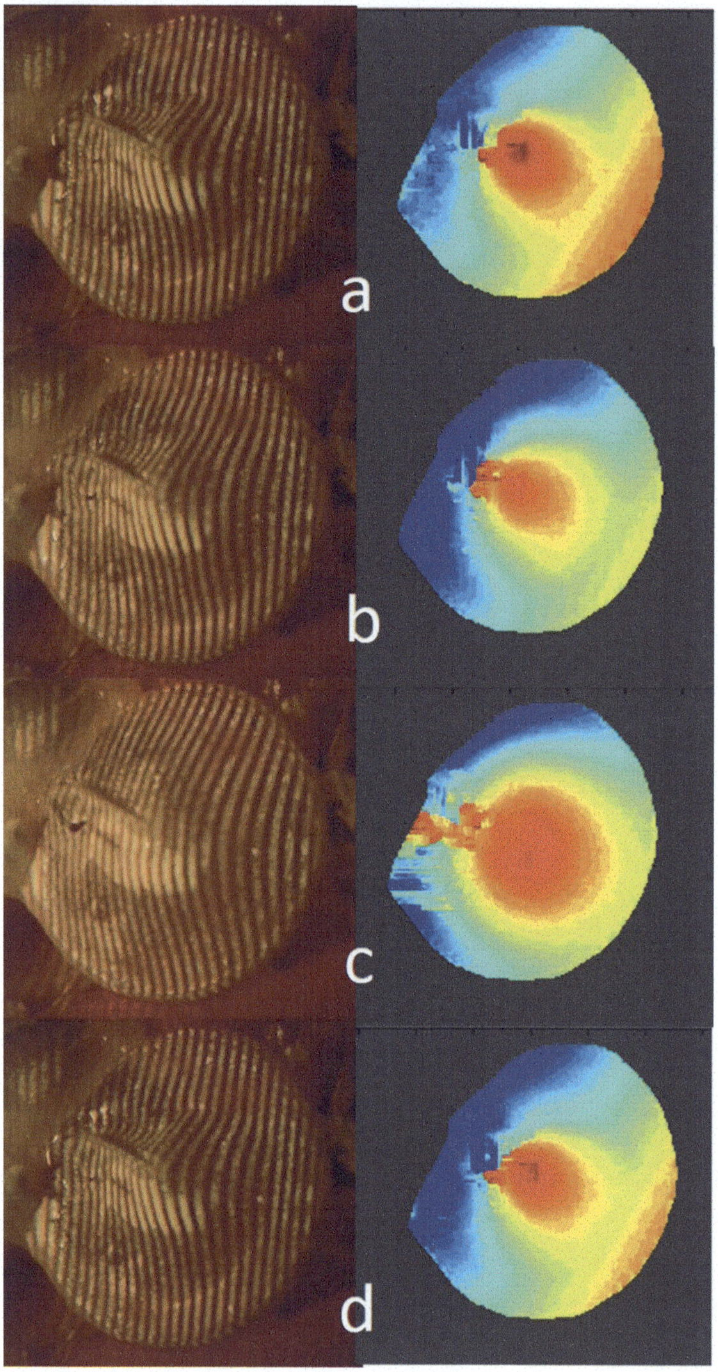

Fig. 7.9 Typical images of deformed chinchilla TM under air pressure at different time: (**a**) t = 0 s, p = 0 kPa; (**b**) t = 0.5 s, p = 1.03 kPa; (**c**) t = 1 s, p = 1.38 kPa; (**d**) t = 1.5 s, p = 0.61 kPa. The color maps demonstrate the 3D-reconstruction of the chinchilla TM

was applied as pulse within 3 s time span. The strain rate was higher than the tensile test. Despite this difference, the results from the two measurements were nearly identical.

Dynamic micro-fringe projection was also used to measure nine chinchilla TMs. It can be noted that, the loadings on the TMs appeared random, though they were all in forms of pulses. In the current setup, the pressures were only applied on the TM without a precise control. The TM under such a form of pressure were usually ruptured. Thus the pressures dropped down when the leakage occurred. The TM material did exhibit viscoelastic behavior to certain extend. However, the overall mechanical performance of TM was dominated by the hyperelastic effect in such a short period. In this case, even a two-term Prony series was able to describe the viscoelasticity behavior for the TM. The current loading is similar to

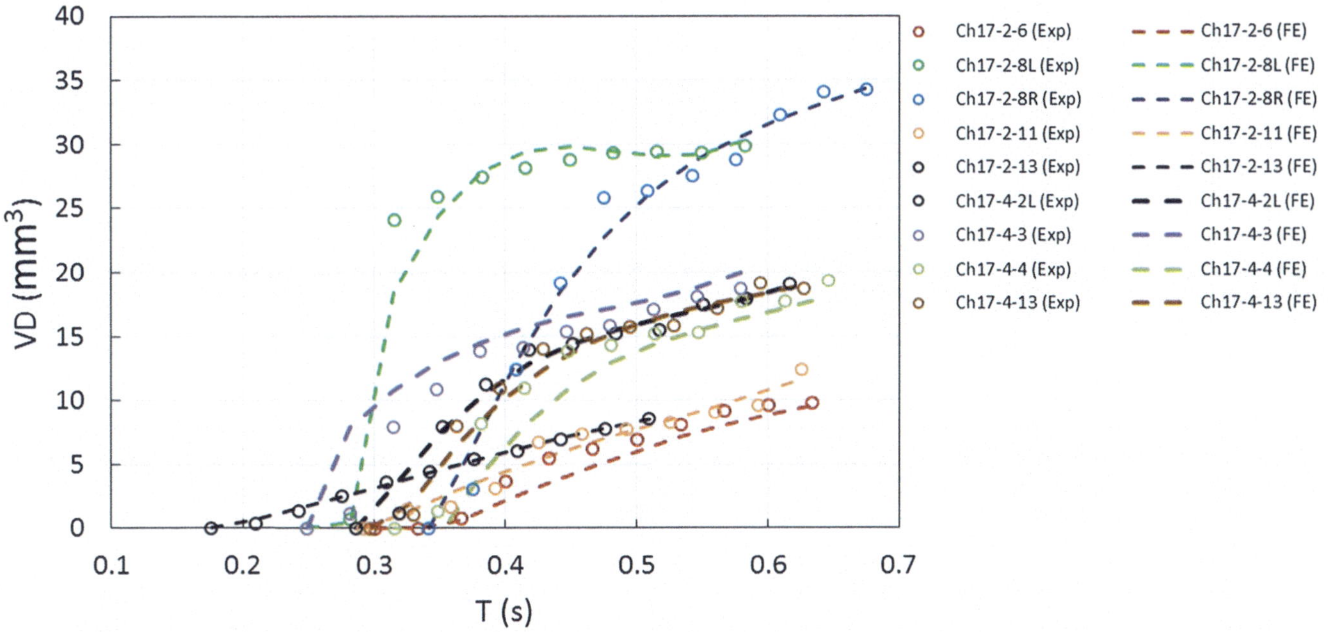

Fig. 7.10 Volume displacements of the five chinchilla TM specimen over time. The circle plots are the experimental data while the dash lines are the FEM simulation results

the blast overpressure applied on the TM but with a relatively longer time span. In the future, more precise control of the pressure (such as creep or periodic pressure loading) will be included in the system to measure the long term effect from the viscoelasticity of the TM.

Acknowledgement We acknowledge the support of DoD W81XWH-14-1-0228 and NSF CMMI1661246.

References

1. Gan, R.Z., Feng, B., Sun, Q.: Three-dimensional finite element modeling of human ear for sound transmission. Ann. Biomed. Eng. **32**(6), 847–859 (2004)
2. Gan, R.Z., Cheng, T., Dai, C., Yang, F.: Finite element modeling of sound transmission with perforations of tympanic membrane. J. Acoust. Soc. Am. **126**(1), 243–353 (2009)
3. Daniel, S.J., Funnell, W.R.J., Zeitouni, A.G., Melvin, D., Schloss, M.D.: Clinical application of finite element model of the human middle ear. J. Otolaryngol. **30**(6), 340–346 (2001)
4. Cheng, T., Dai, C., Gan, R.Z.: Viscoelastic properties of human tympanic membrane. Ann. Biomed. Eng. **35**(2), 305–314 (2006)
5. Zhang, X., Gan, R.Z.: Dynamic properties of human tympanic membrane based on frequency-temperature superposition. Ann. Biomed. Eng. **41**(1), 205–214 (2012)
6. Yokell, Z., Wang, X., Gan, R.Z.: Dynamic properties of tympanic membrane in a chinchilla otitis media model measured with acoustic loading. J. Biomech. Eng. **137**(8), 61–69 (2015)
7. Luo, H., Dai, C., Gan, R.Z., Lu, H.: Measurement of young's modulus of human tympanic membrane at high strain rates. J. Biomech. Eng. **131**(6), 064501 (2009)
8. Daphalapurkar, N.P., Dai, C., Gan, R.Z., Lu, H.: Characterization of the linearly viscoelastic behavior of human tympanic membrane by nanoindentation. J. Mech. Behav. Biomed. Mater. **2**(1), 82–92 (2009)
9. Rohani, S.A., Ghomashchi, S., Agrawal, S.K., Ladak, H.M.: Estimation of the young's modulus of the human pars tensa using in-situ pressurization and inverse finite-element analysis. Hear. Res. **345**, 69–78 (2017)
10. Liang, J., Fu, B., Luo, H., Nakmali, D., Gan, R.Z., Lu, H.: Characterization the nonlinear elastic behavior of guinea pig tympanic membrane using micro-fringe projection. Int. J. Exp. Comput. Biomech. **3**(4), 319–344 (2015)
11. Liang, J., Luo, H., Yokell, Z., Nakmali, D., Gan, R.Z., Lu, H.: Characterization of the nonlinear elastic behavior of chinchilla tympanic membrane using micro-fringe projection. Hear. Res. **339**, 1–11 (2016)
12. Aernouts, J., Dirckx, J.J.J.: Static versus dynamic gerbil tympanic membrane elasticity: derivation of the complex modulus. Biomech. Model Mechanobiol. **11**, 829–840 (2012)
13. Aernouts, J., Dirckx, J.J.J.: Viscoelastic properties of gerbil tympanic membrane at very low frequencies. J. Biomech. **45**, 919–924 (2012)

14. Aernouts, J., Aerts, J.R.M., Dirckx, J.J.J.: Mechanical properties of human tympanic membrane in the quasi-static regime from in situ point indentation measurements. Hear. Res. **290**, 45–54 (2012)
15. Ghadarghadar, N., Agrawal, S.K., Samani, A., Ladak, H.M.: Estimation of the quasi-static Young's modulus of the eardrum using a pressurization technique. Comput. Methods Programs Biomed. **110**, 231–239 (2013)
16. Simo, J.C.: On fully three-dimensional finite strain viscoelastic damage model: formulation and computational aspects. Comput. Methods Appl. Mech. Eng. **60**(2), 153–173 (1987)
17. Holzpfel, G.A.: On large strain viscoelasticity: continuum formulation and finite element applications to elastomeric structures. Int. J. Numer. Methods Eng. **39**(22), 3903–3926 (1996)

Hongbing Lu is a professor, Associate Department Head and Louis A. Beecherl Jr. Chair in the Department of Mechanical Engineering in the Erik Jonsson School of Engineering and Computer Science at the University of Texas at Dallas.

Chapter 8
Realization and Dynamic Studies of CNTs-PDMS Membranes for Biomimetic Flapping Wing Applications

David Kumar, G. M. Kamath, P. M. Mohite, and Sudhir Kamle

Abstract Aerial and aquatic animals including bats, insects and fish use their wings/fins to generate propulsive forces. Natural fliers deform their wings, actively and/or passively, in bending and twisting modes to generate lift and thrust. Within a flapping cycle, wing skin interacts with surrounding fluid and transfers dynamic loads to the internal stiffening structure. Biomimicking of such complex natural flapping wings is possible if the development involves both materials and structural aspects. In the present study, thin PDMS films are chosen for developing the skin of the biomimetic flapping wings. The films are first characterized for dynamic mechanical properties (storage modulus, loss modulus and loss factor) using a dynamic mechanical analyzer. The tests are done in frequency and strain sweep modes to analyze the effect of strain-rates and strain-amplitudes on the dynamic mechanical properties and generate experimental data for constitutive modeling. The dragonfly and cicada wings are taken as the bioinspiration for developing the biomimetic wings. The fabrication of wing skeletons and their integration with the PDMS membranes are achieved through advanced manufacturing techniques including laser micromachining, photolithography and casting. Two types of composite materials are used for making the wing skeletons, i.e., carbon nanotubes (CNTs)/polypropylene (PP) nanocomposite sheet for cicada inspired wing and carbon fiber/epoxy composite strands for dragonfly inspired wing. Structural dynamic analysis of such light, flexible and small size biomimetic structures are interesting and useful for evaluation of biomimicking performance of used materials and manufacturing methods, but difficult to perform. A real-time high-speed non-contact dynamic testing method based on DIC-FPGA coupling (3D digital image correlation technique coupled with real-time data acquisition system, developed at our lab) is used for determining the natural frequencies and corresponding mode shapes of fabricated wings.

Keywords Flapping wings · Soft materials · Nanocomposites · Biomimicking

Aerial and aquatic animals (including bats, insects and fish) use their wings/fins to generate propulsive forces. They deform the wings/fins, actively and/or passively, in bending and twisting modes to generate both lift and thrust. Biomimicking of such structures is possible if the design involves both materials and structural aspects. Within a flapping cycle, the wing skin interacts with surrounding fluid and transfers dynamic loads to internal load bearing structure. The study of a membrane structure is thus useful in designing a biomimetic flapping wing. Flexible polymeric materials including soft materials are useful for developing artificial tissues (such as skin, facial implants, finger joints, heart valves, blood vessels, etc.) and biomimetic structures (such as flapping wings) [1–3]. Carbon nanotube/polypropylene and carbon nanotube/epoxy nanocomposite materials have been developed and characterized for developing biomimetic flapping wings [4–6]. In the present study, thin PDMS membranes are studied with the objective of designing the biomimetic wings.

The PDMS films (Sylgard 184) are fabricated by mixing base solution with the curing agent in 10:1 ratio. The mixture is vigorously stirred manually for 15–20 min and degassed in a vacuum desiccator for 30 min. The curing of the solution is done in an oven for 35 min at 100 °C. The developed PDMS films are characterized for their dynamic mechanical properties using dynamic mechanical analyzer (DMA). MetraVib^{+} 100 DMA is used for this characterization. The dynamic testing in tensile mode is done at different frequencies and dynamic strains. The amplitude of dynamic strain was varied from 0.1% to 2% with the step of 0.1%. The frequency is varied in steps of 5 Hz from 1 Hz to 180 Hz. All the tests were performed at room temperature (26 ± 0.5 °C). The ratio of static to dynamic strain amplitudes is kept at two during the testing. The tests were run in frequency-strain sweep mode. The thickness, width and length of the specimen were 2.74 mm, 10.2 mm and 21 mm, respectively. The storage modulus and loss factor with respect to frequency and dynamic strain amplitudes are shown in Fig. 8.1a, b. It can be seen that the curves for all the dynamic strain amplitude cases are close to each other. For the considered

D. Kumar (✉) · G. M. Kamath · P. M. Mohite · S. Kamle
Department of Aerospace Engineering, Indian Institute of Technology, Kanpur, India
e-mail: david@iitk.ac.in

M. Grady et al. (eds.), *Mechanics of Biological Systems & Micro-and Nanomechanics, Volume 4*, Conference Proceedings of the Society for Experimental Mechanics Series, https://doi.org/10.1007/978-3-319-95062-4_8

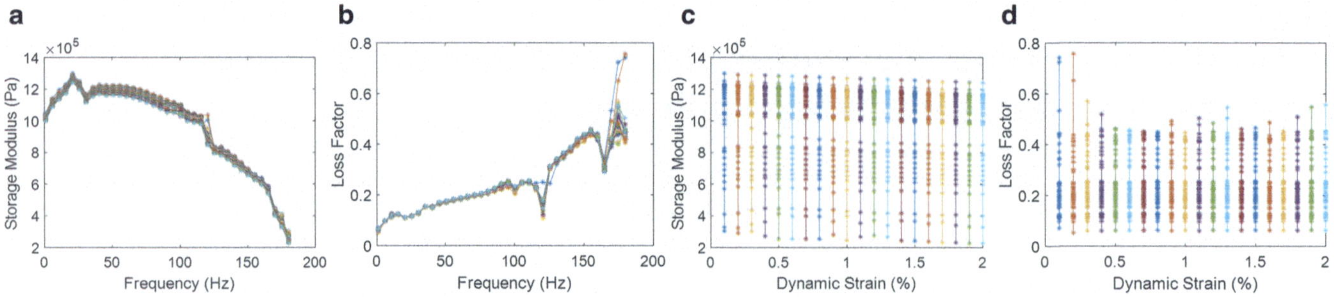

Fig. 8.1 Dynamic mechanical properties of PDMS with respect to frequency and dynamic strain amplitudes

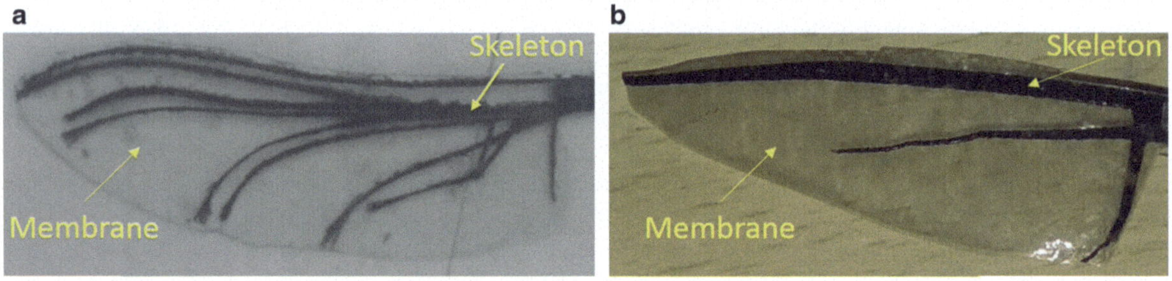

Fig. 8.2 Fabricated PDMS membrane wings, (**a**) Dragonfly inspired wing with carbon fiber/epoxy composite skeleton, (**b**) Cicada inspired wing with CNTs/PP nanocomposite skeleton

frequency range, the storage modulus first increases till 21 Hz and then starts falling down till the end step. In the case of loss factor, it increases throughout the whole range. The unusual jumps in the data can be attributed to instrumental error and/or resonance conditions. The comparison of storage modulus with dynamic strain shows continuous linear decrement (see Fig. 8.1c). On the other hand, loss factor values do not exhibit a similar trend (see Fig. 8.1d). From these results and testing conditions, it can be concluded that the effect of frequency variation is more than the strain amplitude variation. This confirms the strain-rate dependency behavior of PDMS.

Natural fliers flap their wings at resonance to achieve maximum efficiency [7]. At resonance the flapping amplitude becomes high in comparison to off-resonance case, which eventually increases the aerodynamic forces. Therefore, an artificial wing should be flapped close to its natural or fundamental frequency. Combes and Daniel [8] found that the surrounding air does not change the structural deformations of a natural wing during resonant flapping cycle. Biomimetic wing typically comprises a soft membrane with a stiffener skeleton. In the present study, fabrication of wing skeletons and their integration with membrane are achieved through advanced manufacturing techniques including laser micromachining, photolithography and casting. Two types of composite materials are used for making the skeletons i.e., carbon fiber/epoxy composite and CNTs/PP nanocomposite. The skeleton out of carbon fiber/epoxy composite is fabricated using a PDMS mold. The design of this mold is inspired by dragonfly wing. The length and chord of dragonfly inspired wing are 44.66 mm and 15.09 mm, respectively. The manufacturing of mold is done using photolithography. On the other hand, the laser micromachining is used to cut the skeleton from CNTS/PP nanocomposite sheet. The cicada wing is chosen as the inspiration for CNTs/PP nanocomposite skeleton. The cicada inspired wing skeleton was over simplified due to power and resolution limitations of used laser micromachining machine. The length and chord of simplified cicada inspired wing are 50 mm and 15.6 mm respectively. The skeletons are embedded in the PDMS membrane through the casting process. The fabricated wings are shown in Fig. 8.2.

Structural dynamic testing of such light, flexible and hyperviscoelastic structures is useful but difficult to perform. This work proposes an experimental technique to characterize dynamic properties of soft materials and structures. A high-speed 3D digital image correlation (DIC) system is developed for obtaining the dynamic displacements/deformations. Two high-speed cameras are set in stereo-vision configuration for recording the images of the vibrating beams. An in-house MATLAB code is developed to correlate the images and determine the full-field displacements. The DIC technique provides only deformations/displacements of an object moving in 3D space. However for performing structural dynamic analysis, the object has to be excited and the corresponding response has to be measured along with input. One of the main requirements for such experiments is that the input and output has to be properly synchronized i.e. there should not be any time-lag between the excitation and the response. The real-time operating system with embedded FPGA (RT-FPGA) is used to couple

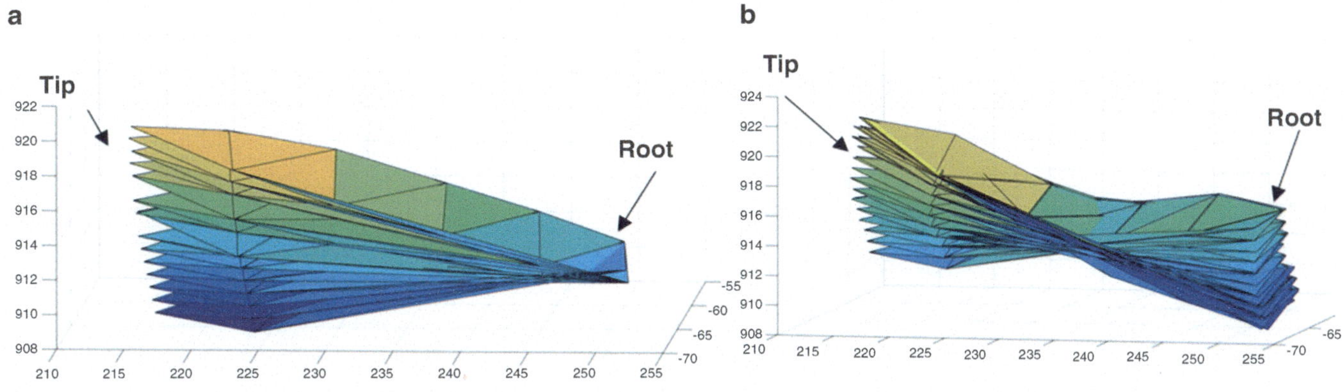

Fig. 8.3 Mode shapes of the dragonfly inspired wing with carbon fiber/epoxy composite skeleton, (**a**) Mode-1: Flexural mode, (**b**) Mode-2: Flexural mode with a node in the middle of the wingspan

the DIC system with other sensors. The cameras are coupled and synchronized with accelerometer and laser displacement sensor using RT-FPGA. This technique is validated by performing structural dynamic testing of ASTM-standard beams [6]. This real-time high-speed non-contact DIC-FPGA setup has been used for structural dynamic analysis of light and flexible structures such as membranes and flapping wings [9]. The natural frequencies of wings are obtained by performing FFT of measured displacement data. Images from cameras are also correlated to determine mode shapes. Wings are speckle painted to make the patterns and perform correlations over images recorded during vibration cycles. The first and second natural frequencies of dragonfly inspired wing (made of carbon fiber/epoxy skeleton and PDMS membrane) are 11 Hz and 15 Hz, respectively. This wing has flexural mode shapes, as shown in Fig. 8.3. Similar mode shapes are found in literature.

In the future of this study, the wings with CNTs/PP nanocomposite skeletons will be developed with different skeleton configurations and tested for performing optimization exercise towards the biomimicking goal. These wings will also be compared with the wings of carbon fiber/epoxy composite skeletons. Similar studies will also be carried out computationally using Abaqus and validated with the experimental results. Hyperviscoelastic modeling of PDMS will be done for performing finite element simulations. The experimental measurements will be performed inside a vacuum chamber to observe the effect of air and compare with computational results. The piezoelectric properties of CNTs could potentially be exploited for sensing and control for morphing and flapping wing applications.

References

1. Lin, I.K., et al.: Viscoelastic mechanical behavior of soft microcantilever-based force sensors. Appl. Phys. Lett. **93**(25), 15–18 (2008)
2. Bentil, S.A., Ramesh, K.T., Nguyen, T.D.: A dynamic inflation test for soft materials. Exp. Mech. **56**(5), 759–769 (2016)
3. Chakravarty, U.K.: Analytical and finite element modal analysis of a hyperelastic membrane for micro air vehicle wings. J. Vib. Acoust. **135**(5), 1–6 (2013)
4. Kumar, D., et al.: Development and analysis of aeroelastically tailored carbon fiber-CNT/PP composite bioinspired wings. In: 33rd AIAA Applied Aerodynamics Conference, AIAA AVIATION Forum (2015)
5. Kumar, D., et al.: Development and modal analysis of bioinspired CNT/epoxy nanocomposite MAV flapping wings. J. Aerosp. Sci. Technol. **67**(1), 88–93 (2015)
6. Kumar, D., et al.: Dynamic studies of flexible sandwich beams with hyperviscoelastic Core using digital image correlation. In: International Conference on Composite Materials and Structures (2017)
7. Greenewalt, C.H.: The wings of insects and birds as mechanical oscillators. Proc. Am. Philos. Soc. **104**(6), 605–611 (1960)
8. Combes, S.A., Daniel, T.L.: Into thin air: contributions of aerodynamic and inertial-elastic forces to wing bending in the hawkmoth Manduca sexta. J. Exp. Biol. **206**(17), 2999–3006 (2003)
9. Kumar, D., et al.: A novel structural dynamic testing method for light and flexible structures possessing material and structural complexities. Manuscript under preparation

David Kumar is a Sr. Project Engineer at Department of Aerospace Engineering, Indian Institute of Technology Kanpur, India. He did his Masters in 2011, and Ph.D. in 2018 from the same institute.

Chapter 9
Experimental and Theoretical Study on the Robustification of Acoustic Emission Inspection with Recurrent Neural Networks

Christopher Hsu, Mulugeta Haile, Natasha Bradley, and John Chen

Abstract This paper presents a summary of an experimental and theoretical study conducted to improve the performance of Acoustic Emission (AE) based structural health monitoring. Two broadband AE sensors are attached to a test specimen and a series of artificial test signals were generated using multiple pencil-lead breaks. The geometry of the test specimen is such that; one of the AE sensors always receives a direct wave and the second sensor receives reflection (or echo) of the same direct wave received at the first sensor. Time series data of the direct and reflected waves, recorded at a rate of 5 MHz, are used to train a multi-layer recurrent neural network (RNN) in a supervised setting. The inherent patterns learned from the training data are used to classify received acoustic emission waveforms into reflection and direct waves. Preliminary results show that the proposed technique, at least in a similar setting and material, improves the performance of AE based health monitoring.

Keywords Acoustic emission · Recurrent neural network · LSTM · Classification · Sequence

9.1 Introduction

The majority of existing work on acoustic emission-based structural health monitoring is for geometrically simple structures with uninterrupted propagation paths. Realistic systems such as a full-scale aircraft, however, are built from interconnected components such as sandwich core panels, stringer stiffened skins, and fastener holes. The geometric complexity of realistic systems causes acoustic emission waves to undergo multiple reflections, refractions, and mode changes [1–3]. Conventional techniques for identification of AE wave reflections rely on acquisition time control parameters. In this approach, a small time period, typically in microseconds, is specified as the minimum time that separates consecutive direct hits. As such, if an AE wave arrives at the same transducer location before this time period has elapsed then the wave is considered a reflection and is excluded from further analysis. This method, despite being acceptable for geometries where reflection boundaries are sparse and well separated, does not work well with complex airframe systems where multiple reflection boundaries are tightly distributed in close proximity to the sensor and source locations. Conventional signal processing techniques using traditional AE signal statistics, such as the rise-time, duration, signal energy, or transform domain characteristics such as wavelet parameters, do not work since the direct wave and its reflection have almost identical properties. In this work, recurrent neural networks (RNN) are used to learn distinct patterns of the direct and reflected AE waves in much higher dimension. Recurrent nets are a type of artificial neural network well suited to recognize patterns in time series data such as the type generated by AE sensors. Long Short-Term Memory (LSTM) is a variation of RNN with certain computational advantage to the challenge addressed in this paper.

Long Short-Term Memory (LSTM) has been around for quite some time. Introduced by Hochreiter and Schmidhuber [4], they have been used in the pursuit of sequence learning. These models have been used for sequence classification [5] as well as sequence prediction [6]. Recent LSTMs similarly resemble the older models except with the inclusion of forget gates [7] and peep-hole connections [8]. The main reason to use the LSTM structure over the other RNN methods is to avoid the vanishing gradient problem. In this paper, the model reflects the use of the forget gates and sequence classification.

C. Hsu (✉) · M. Haile · N. Bradley · J. Chen
U.S. Army Research Laboratory, Vehicle Technology Directorate, Aberdeen Proving Ground, MD, USA
e-mail: christopher.d.hsu.ctr@mail.mil

M. Grady et al. (eds.), *Mechanics of Biological Systems & Micro-and Nanomechanics, Volume 4*, Conference Proceedings of the Society for Experimental Mechanics Series, https://doi.org/10.1007/978-3-319-95062-4_9

9.2 Materials and Methods

Two broadband acoustic emission transducer (Model KRNm300) are attached to each side of the U-shaped aluminum plate shown in Fig. 9.1. These transducers are electrodeless nickel plated brass constructed with an electrically isolating ceramic face. The transducers have a sensitivity range of 200–850 kHz and are connected to a pre-amplifier which has a gain of 34 dB (Model AEP4). Acoustic emission data is acquired and stored using a high speed data acquisition hardware (Model AMSY-6). The acquisition threshold is set at 40 dB, with a sampling rate of 5 MHz, and the pre- and post- trigger points are set at 200. The U-shaped aluminum plate, shown in Fig. 9.1, is designed in such a way that each pencil break will produce a direct wave for the corresponding sensor and a reflection wave on the opposite sensor. For example, if a pencil is broken at location R3, channel 1 will receive a direct propagated wave while channel 2 will receive a reflection wave as there is no linear direct path for the wave to travel. This test setup creates an equal number of direct and reflection waves that can then be used for training and testing. The artificial acoustic emission generation follows the standard Hsu-Nielson practice, where a 0.5 mm lead pencil with a rubber guard is held at a 45° angle and broken. Pencil breaks are conducted at the specified locations indicated in Fig. 9.1 below. Each test experiment includes 21 breaks which produce 42 samples of direct and indirect waves. Future tests will include random breaks closer to the boundaries of the aluminum plate rather than just in the center in order to vary the training data set.

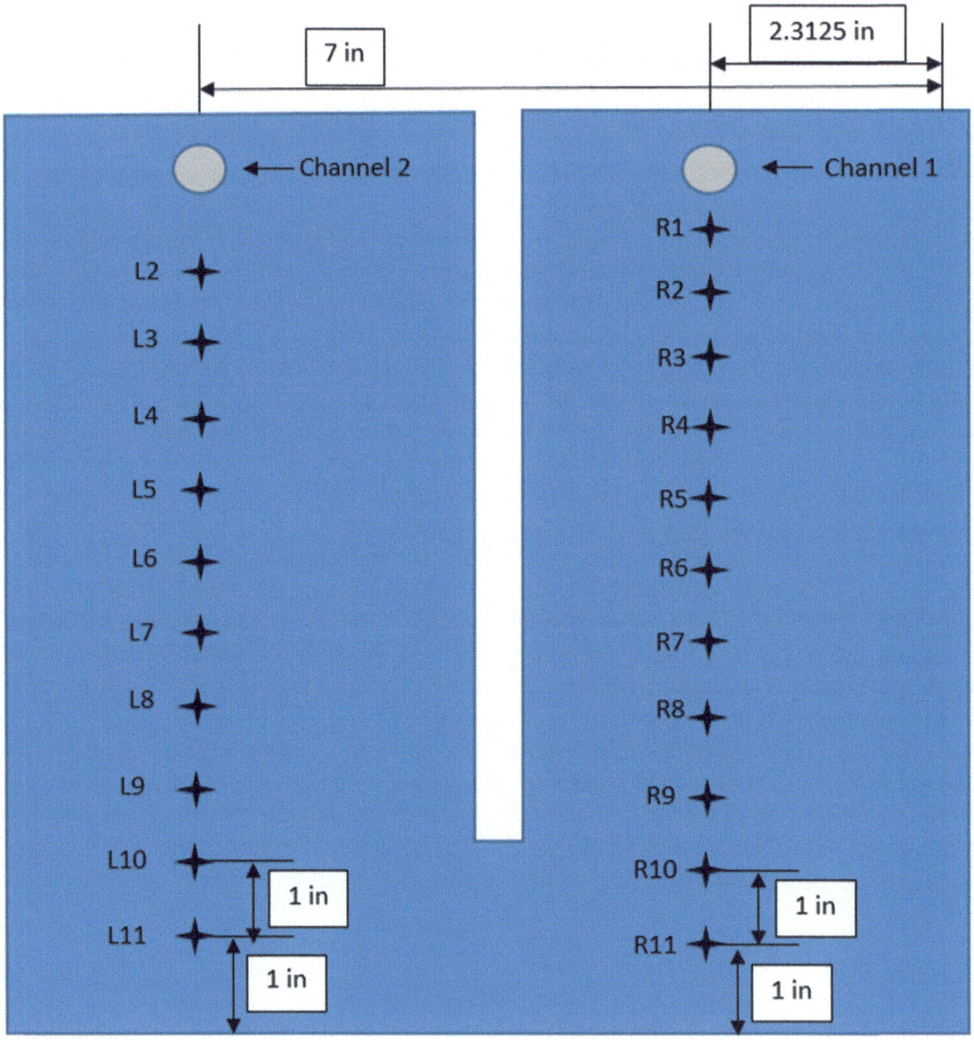

Fig. 9.1 The test specimen is a grooved (U-shaped) aluminum sheet with locations. Two acoustic emission transducers, channel 1 and 2 are attached at the locations shown. Location of the Hsu-Nielson (or Pencil Lead Break) sources are marked as L2–L11 for the left half of the grooved plate and R1–R11 for the right

9.3 Data Pre-processing

In this work, we are looking to correctly identify whether a recorded acoustic wave is direct wave (direct line of sight between source and channel) or reflected wave. Being able to accurately distinguish between the two can have significant benefits on the accuracy of future AE-based structural damage detection and localization. The classification of the acoustic emission wave is characterized as a binary output, 0 for direct and 1 for reflection. Labeled data is collected as described above with each pencil break producing a direct wave at one channel location and reflections at the other channel location. In order for the data to be processed in the LSTM network, it needs to go through some preprocessing steps. Since LSTMs are made for long term dependencies, computational memory usage can rise significantly for long (length) time-series datasets. Due to the format of data collection in the test, some samples can record up to 64,000 data points, which is an extremely large amount for our purposes. To process this many data points for each sample in a classical neural network, it would take heavy computing power which is just not feasible in the goal of putting this technology for on-board application. In general, it is beneficial to resize the data by removing points that do not add value to the training of the network, such as the pre and post trigger points as well as prolonged near threshold data points. This process improves the computational efficiency and accuracy of the network. By limiting the data to what is being identified, the network is much less likely to over fit or even skip over the essential attributes.

For each experiment (42 signal samples), each sample is shortened by down sampling the signal to 2.5 MHz (from the original 5 MHz). Hence, the data is then shortened to 2000 data points which include 100 pre-triggered points. An example of the original data and the truncated data for both the direct and reflected waves are shown in Fig. 9.2.

9.4 Results

All datasets are concatenated together and then split up and randomized with 80% used for training and validation while the remaining 20% is used for testing. We adjusted the network parameters to best suit this dataset. In the end, we trained the LSTM with 256 epochs. Two stacked LSTM layers are used with 64 hidden nodes in each layer. The L2 regularization term is used to avoid exploding gradients and the Adam optimizer is used to combat the noise problem which is common in sensor data. We also use a batch size of 64, a learning rate of 0.0015 and a lambda loss amount of 0.0015 for the L2 term to train the system. When training is initiated, the data is quickly evaluated to check if it has been properly normalized. When that is confirmed, the training begins. Throughout the training procedure, at each epoch, the model is evaluated by the training set and is also tested against the test set. Accuracies and batch losses are calculated for training and is shown in Fig. 9.3. When all the training is complete, the model is compared to the test set followed by calculation of performance metrics. Performance metrics used in this test are precision, recall and F1 score (Table 9.1 and Fig. 9.4).

9.5 Discussions

It can be seen from the results that LSTM RNNs, can successfully classify direct and reflection acoustic emission waves. Here, the LSTM was able to classify a test set with an accuracy over 90%. The same model had a 100% training set accuracy. However, there is still room for the model to be improved. While those figures are great for this specific situation, the test data still stems from data collected in a lab where most factors can be controlled, i.e., such as constant temperature, minimal ambient noise, and relatively simple structural geometries. While our data sets can be seen as large, it is small relative to datasets found in other deep learning applications such as vision and speech recognition. Those datasets have millions of variations that can give the classifier a chance to learn every possible scenarios. The reason to mention this is to state the importance of generalization and the problem of overfitting. The LSTM network's goal is to optimize its weights and biases based on the loss from the training set. It can be seen in the training log that over several epochs, training accuracies increase for the test set and slowly reach a perfect accuracy of 100%. However, this success can negatively affect the test accuracy due to overfitting.

Overfitting is a common problem in deep learning applications. If trained too specifically for a given training set, the model will not generalize well to new data. There are a few techniques in order to generalize a model which include more training samples of different scenarios, use of dropout and randomization at each epoch. In the future, we will look to create more data with different scenarios such as performing pencil break at non-centered locations on the test specimen. This

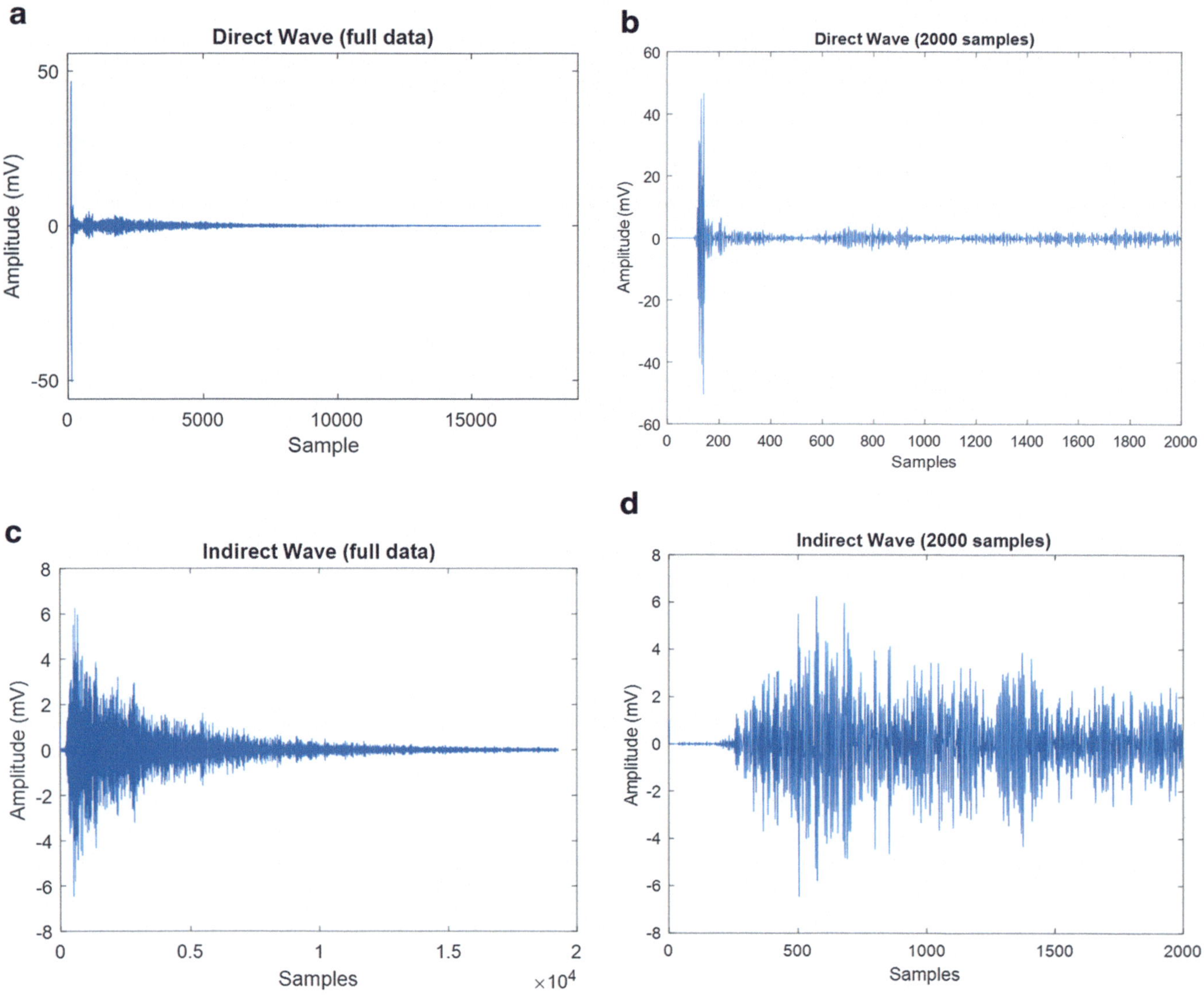

Fig. 9.2 Direct and reflection waves propagated from Hsu-Nielson test located at point R1. (**a**) Direct wave full length. (**b**) Direct wave shortened to 2000 samples. (**c**) Reflection wave full length. (**d**) Reflection shortened to 2000 samples

process could diversify the training data by varying timing and reflection angles. Different sizes and shapes of aluminum plates can also be used with additional nuts and bolts put in place in order to simulate a more realistic airframe structure. Besides combatting overfitting through increased data size and diversity, the implementation of dropout and randomization at epochs can be used. Dropout is the technique of randomly ignoring certain nodes in the network in order to create some disorder in the data. Training has been seen to create dependencies amongst normally connected nodes, but by ignoring a few based on a certain probability, we can hope to dissuade the creation of these dependencies which produce an overfitted model. Finally, we can also look to randomize the data at every epoch. In this paper, the data is randomized at the very beginning but from there on out it stays in that order. Similar to the idea of dropout, the specific order can create some over fit features between samples of the data. Through the use of randomization at every epoch, the network will be fed a new combination of these same samples to ensure there is no correlation in the order.

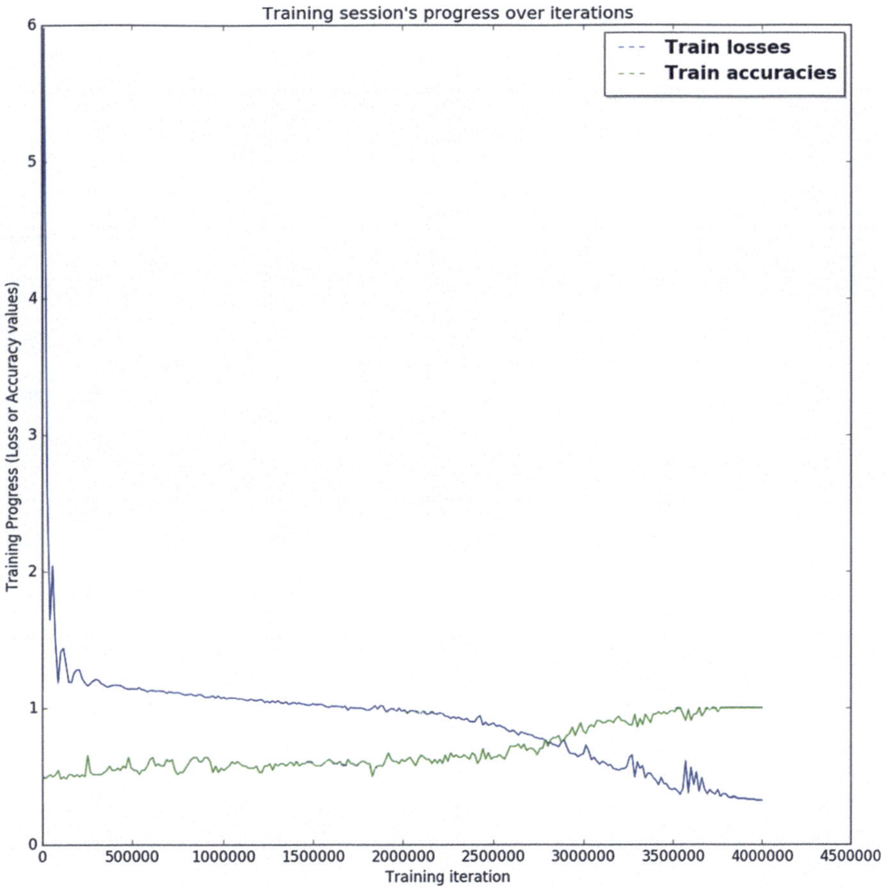

Fig. 9.3 Training losses and accuracies over training iterations for acoustic emission data obtained using two broadband sensors on 2D plate

Table 9.1 Performance results and metrics

Performance metrics	Value
Final training loss	0.323853
Final training accuracy	1.0
Final testing loss	0.7695973515510559
Final testing accuracy	0.9411764740943909
Precision	0.9470588235294119
Recall	0.9411764705882352
F1 score	0.9407636738906089

9.6 Conclusions

All in all, this project has shown that LSTM RNN's can be successfully used to improve the performance of acoustic emission based structural health monitoring by accurately classifying direct and reflected waves in geometrically complex structures. This paper is a summary of an attempt to automatically discern the difference between a direct wave and a reflection wave. While our LSTM produces promising results, this is only the first step in building a full diagnostic system. The purpose for using deep learning techniques is to be able to use raw data. The fact that it was necessary to preprocess the data is not ideal for real world application. Although it seems clear that only the first portions of the data samples are important for classification, it is hard to fully discount that the rest of the data is necessary. Also, the need for the preprocessing steps presents an inefficiency from data collection to classification. In future projects, we will focus on the elimination of the preprocessing step. Finally, we will look to implement the other techniques stated above in order to generalize our model. Overfitting is a common problem in deep learning and it is necessary to take measures to prevent it. The robustness of our model can be further improved by generating diverse sets of training data that mimics all possible scenarios of AE sources, reflection interfaces, material conditions as well as test environment.

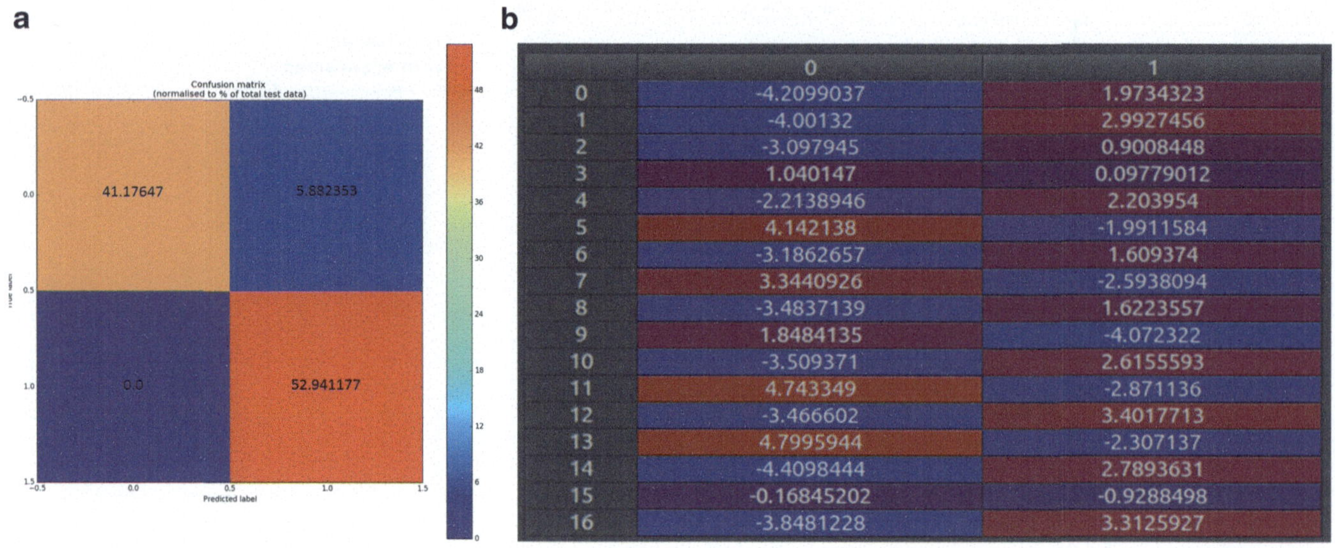

	0	1
0	-4.2099037	1.9734323
1	-4.00132	2.9927456
2	-3.097945	0.9008448
3	1.040147	0.09779012
4	-2.2138946	2.203954
5	4.142138	-1.9911584
6	-3.1862657	1.609374
7	3.3440926	-2.5938094
8	-3.4837139	1.6223557
9	1.8484135	-4.072322
10	-3.509371	2.6155593
11	4.743349	-2.871136
12	-3.466602	3.4017713
13	4.7995944	-2.307137
14	-4.4098444	2.7893631
15	-0.16845202	-0.9288498
16	-3.8481228	3.3125927

Fig. 9.4 Confusion matrix of the LSTM network after the final epoch. (**a**) Confusion matrix of test data. (**b**) matrix of test predictions (red indicating likely class)

References

1. Haile, M., Chen, T.K., Sediles, F., Shiao, M., Le, D.: Estimating crack growth in rotorcraft structures subjected to mission load spectrum. Int. J. Fatigue. **43**, 142–149 (2012)
2. Haile, M.A., Bordick, N.E., Riddick, J.C.: Distributed acoustic emission sensing for large complex air structures. J. Struct. Health Monit. **17**, 624–634 (2017). https://doi.org/10.1177/1475921717714614
3. Haile, M.A., Riddick, J.C., Assefa, A.H.: Robust particle filters for fatigue crack growth estimation in rotorcraft structures. IEEE Trans. Reliab. **65**(3), 1438–1448 (2016)
4. Hochreiter, S., Schmidhuber, J.: Long short-term memory. Neural Comput. **9**(8), 1735–1780 (1997)
5. Elman, J.L.: Finding structure in time. Cogn. Sci. **14**(2), 179–211 (1990)
6. Bengio, S., Vinyals, O., Jaitly, N., Shazeer, N.M.: Scheduled sampling for sequence prediction with recurrent neural networks. In: Advances in Neural Information Processing Systems 28 (NIPS 2015)
7. Gers, F.A., Schmidhuber, J., Cummins, F.: Learning to forget: continual prediction with LSTM. Neural Comput. **12**(10), 2451–2471 (2000)
8. Gers, F., Schmidhuber, J.: Recurrent nets that time and count. In: Proceedings of the IEEE-INNS-ENNS International Joint Conference on Neural Networks (IJCNN), vol. 3, pp. 189–194. IEEE (2000)

Christopher Hsu Journeyman Fellow at the US Army Research Laboratory, Vehicle Technology Directorate. As a fellow in the Mechanics Division, Probabilistic Risk Assessment Team, I look to apply machine/deep learning techniques for airframe structural analysis.

Chapter 10
Modeling of Atomic Force Microscope Contact Experiments on Escherichia coli Bacteria Cellular Systems

Daniel Biggs, Hanwei Liu, David Tirrell, and Guruswami Ravichandran

Abstract We examined common models for atomic force microscope (AFM) nanoindentation of living cells and propose a new methodology for fitting the mechanical properties to the indentation data. By modeling the living cells as fluid-filled spherical membranes, the important parameters become the cell wall thickness, cell wall elastic modulus, and the cell's turgor pressure. We modeled these parameters in addition to adhesion, plasticity, and viscoelasticity using finite element analysis to better describe the underlying physics and geometry of microbial indentation. This modeling approach applied different scaling laws from the classical Hertz model in order to better explain indentation data found in experiments on living E. coli bacterial cells. Work supported by DARPA Contract No. HR0011-17-2-0037.

Keywords Nanoindentation · Atomic force microscope · E. coli bacteria · Finite element analysis · Contact mechanics

Cell processes such as growth, motility, and division are governed by mechanical interactions of cellular structures with their environment and neighboring cells. While indentation of living cells using an Atomic Force Microscope (AFM) has provided a wealth of experimental data, extracting the mechanical properties of cellular structures has been difficult partly due to the large variability inherent in biological systems, but also due to a lack of consensus in the interpretation of the indentation data. A better understanding of the underlying cell physics is needed to provide a more accurate and reliable measure of the mechanical properties of cell structures. Decoupling the impact of cell elasticity, viscoelasticity, adhesion, turgor pressure, and plasticity through nanoindentation experiments would provide additional insights into the basic mechanics of these biological systems.

The most common approach to model the elastic properties of living cells is to either use the Hertz contact model (or some derivative such as the Sneddon model for paraboloid indenters), or a linear spring constant model [1]. The Hertz model provides an accurate solution when its assumption of small indentation of a homogenous solid is well founded, and has been widely used even for cell nanomechanics due to the simplicity of its short formulaic expression and its general acceptance in the broader AFM community. The JKR model, which is an extension of the Hertz model to include effects of adhesion, has also been applied to model AFM based indentation data of soft materials [2]. The viscoelastic properties of living cells modeled as a power-law material have also been successfully measured with an AFM by a Force-Clamp method [3]. Recent works on both plant cells and bacterial cells have begun to relay the importance of modeling these living cells as fluid filled structures with an internal turgor pressure, as opposed to homogenous solids [4, 5]. This provides a more accurate representation of the cellular mechanics, but fitting experimental AFM indentation data to these models has proven to be challenging because there are now multiple fitting variables including the elastic modulus and thickness of the cell wall, and the turgor pressure instead of a single elastic modulus. For specific cells such as E. coli, the cell wall thickness and elasticity have been measured and calculated, and can be inputted as known quantities [6, 7]. Bacterial cells such as E. coli which form biofilm colonies have been shown to exhibit a large amount of adhesion [8], and there have been no models to date that capture both the turgor pressure and adhesion of bacterial cells. Here, we model the turgor pressure and adhesion of E. coli cells using finite element analysis (FEA) and relate it to phenomena witnessed in experimental AFM indentation data.

In this study, indentation experiments on living cells were conducted using an Asylum Research MFP-3D Bio AFM. This AFM has an inverted Olympus IX-51 optical stand and microscope that was used to help locate cells prior to indentation. Silicon nitride cantilevers with glass beads attached were used to indent live E. coli cells both in air and in minimal nutrient

D. Biggs (✉) · G. Ravichandran
Division of Engineering and Applied Science, California Institute of Technology, Pasadena, CA, USA
e-mail: dbiggs@caltech.edu

H. Liu · D. Tirrell
Division of Chemistry and Chemical Engineering, California Institute of Technology, Pasadena, CA, USA

M. Grady et al. (eds.), *Mechanics of Biological Systems & Micro-and Nanomechanics, Volume 4*, Conference Proceedings of the Society for Experimental Mechanics Series, https://doi.org/10.1007/978-3-319-95062-4_10

media. The cantilever stiffness was calibrated prior to indentation using the thermal tune method. The stiffness of the bacteria cells was found to vary greatly when indented in minimal media versus in air. The indentation data exhibited large amounts of hysteresis, indicating possible plastic deformation and viscoelasticity, and large amounts of adhesion on the retraction curve.

The finite element simulations of the nanoindentation experiments were conducted using Abaqus FEA. The problem was formulated using an axisymmetric model with a fluid filled spherical cell resting on a flat rigid substrate and indented by a spherical rigid indenter. The cell had an outer radius of 1.0 μm and a nominal cell wall thickness of 10 nm, although several wall thicknesses were tested. The interior of the cell was modeled as a fluid-filled cavity with constant volume. The Young's modulus of the cell wall was 25 MPa and had a Poisson's ratio of 0.37. The indentation simulation was solved in two steps, first, a thermal strain was induced on the cell wall in order to cause a thermal contraction and build a variable turgor pressure ranging from 0 to 30 KPa, and second, the indenter was displaced into the cell by 200 nm, which is 10% the diameter of the cell. This caused two areas of contact, one between the indenter and the cell, and a second between the flat substrate and the cell. Both were addressed by the general contact algorithm within Abaqus as frictionless and hard contact conditions. The cell wall was meshed using CAX8 elements. Varying amounts of adhesion was modeled using a cohesive interaction property, and plasticity was introduced using a Von Mises criterion with isotropic hardening. The indenter position and reaction force on the indenter were then extracted from the solution output and compared with AFM experimental data.

The simulation results had large differences compared to the classical Hertz contact model, particularly with regards to scaling with indenter radius. The power law for the indentation, which is constant for the Hertz model, was found to vary greatly based on the initial turgor pressure and the indenter radius. The experimental data was compared against the finite element simulations to discern the varying importance of the initial turgor pressure, adhesion, plasticity, and viscoelasticity. This work highlights the importance of the underlying physics and geometry involved in indenting live E. coli cells.

References

1. Gaboriaud, F., Dufrêne, Y.: Atomic force microscopy of microbial cells: application to nanomechanical properties, surface forces and molecular recognition forces. Colloids Surf. B: Biointerfaces. **54**, 1–10 (2006)
2. Notbohm, J., Poon, B., Ravichandran, G.: Analysis of nanoindentation of soft materials with an atomic force microscope. J. Mater. Res. **27**, 229–237 (2011)
3. Hecht, F.M., Rheinlaender, J., Schierbaum, N., Goldmann, W.H., Fabry, B., Schäffer, T.E.: Imaging viscoelastic properties of live cells by AFM: power-law rheology on the nanoscale. Soft Matter. **11**, 4584–4591 (2015)
4. Weber, A., Braybrook, S., Huflejt, M., Mosca, G., Routier-Kierzkowska, A.L., Smith, R.S.: Measuring the mechanical properties of plant cells by combining micro-indentation with osmotic treatments. J. Exp. Bot. **66**(11), 3229–3241 (2015)
5. Deng, Y., Sun, M., Shaevitz, J.W.: Direct measurement of cell wall stress stiffening and turgor pressure in live bacterial cells. Phys. Rev. Lett. **107**, 1–4 (2011)
6. Yao, X., Jericho, M., Pink, D., Beveridge, T.: Thickness and elasticity of gram-negative murein sacculi measured by atomic force microscopy. J. Bacteriol. **181**, 6865–6875 (1999)
7. Boulbitch, A., Quinn, B., Pink, D.: Elasticity of the rod-shaped gram-negative eubacteria. Phys. Rev. Lett. **85**, 5246–5249 (2000)
8. Oh, Y.J., Jo, W., Yang, Y., Park, S.: Influence of culture conditions on Escherichia coli O157:H7 biofilm formation by atomic force microscopy. Ultramicroscopy. **107**, 869–874 (2007))

Daniel Biggs completed his Ph.D. at the University of Michigan researching shape memory alloy cables advised by Prof. John Shaw. He is currently a postdoc at Caltech researching the mechanical properties of living materials with Prof. Ravichandran.

Chapter 11
Identification of Shear Sample Test Geometry for Bulk Adhesive Characterization

Brock Watson, Michael J. Worswick, and Duane S. Cronin

Abstract High-fidelity finite element models of adhesively bonded structures require bulk adhesive mechanical properties describing stress-strain response, dependence on deformation rate, and ultimate strength of the adhesive material. In particular, the shear properties of the adhesive are necessary to accurately represent Mode II and mixed-mode loading response, but are not directly available from typical characterization tests such as the lap shear test. To address this challenge, five proposed shear specimen geometries from the literature were machined from flat plaques of cured two-part toughened structural epoxy and assessed by comparing the stress-strain response for quasi-static loading. The stress was calculated from the measured force and ligament area while strain was measured using Digital Image Correlation. There was generally good correspondence between the stress-strain responses for all specimens, although the strain at failure varied between the different sample geometries. In all cases, failure initiated at locations of high hydrostatic stress, as noted by others when testing polymers under shear loading or areas of high stress concentration. The sample with the smallest gauge length provided the highest strain measurement prior to failure and thus the most complete stress-strain response. However, in view of the failure initiation outside of the ligament, the measured shear strains at failure should be treated as a lower bound. Additionally, the small gauge length is conducive to future high deformation rate testing, since lower test velocities would be needed for high loading rates to be applied to the sample.

Keywords Structural adhesive · Polymeric material characterization · Shear testing · Digital image correlation · Toughened epoxy

11.1 Introduction

The characterization of adhesive response in shear is an important consideration when developing finite element models of adhesively bonded structures. One approach to modeling bonded joints involves using continuum constitutive and failure models of each component (*i.e.,* adherends and adhesive) independently, which requires high quality measurement of the shear stress-strain response. However, the small deformation associated with typically thin bond lines (*e.g.,* 0.17 mm) and the ability to generate pure shear loading present challenges in measuring these properties. A number of studies have investigated the response of glassy polymers in shear loading; however, these studies require special test equipment [1, 2] or torsional loading of the specimen [3, 4]. While equipment does exist to test samples under torsion at high loading rates, it is generally easier to use a sample geometry requiring tension or compression loading at elevated rates. Testing of bulk adhesives under shear loading has been carried out previously using the Iosipescu shear sample geometry, which exhibited failure near locations of high hydrostatic stress [2] and torsion samples that exhibited spiral fractures [4]. Both failure modes indicate a strong dependence on tensile strength to initiate fracture of the samples under shear loading [5, 6].

The aim of the current study was to assess five shear test sample geometries for bulk shear response using a toughened two-part epoxy adhesive subjected to quasi-static loading. Five existing sample geometries developed for metals (Fig. 11.1) were identified; the 'mini-shear' sample [7], a modification of the mini-shear geometry termed the 'micro-shear' sample [8], the 'block-shear' sample [9], the ASTM B831, and the ASTM B831 Alternate [10] test samples.

B. Watson (✉) · M. J. Worswick · D. S. Cronin
Mechanical and Mechatronics Engineering, University of Waterloo, Waterloo, ON, Canada
e-mail: bwatson@uwaterloo.ca

M. Grady et al. (eds.), *Mechanics of Biological Systems & Micro-and Nanomechanics, Volume 4*, Conference Proceedings of the Society for Experimental Mechanics Series, https://doi.org/10.1007/978-3-319-95062-4_11

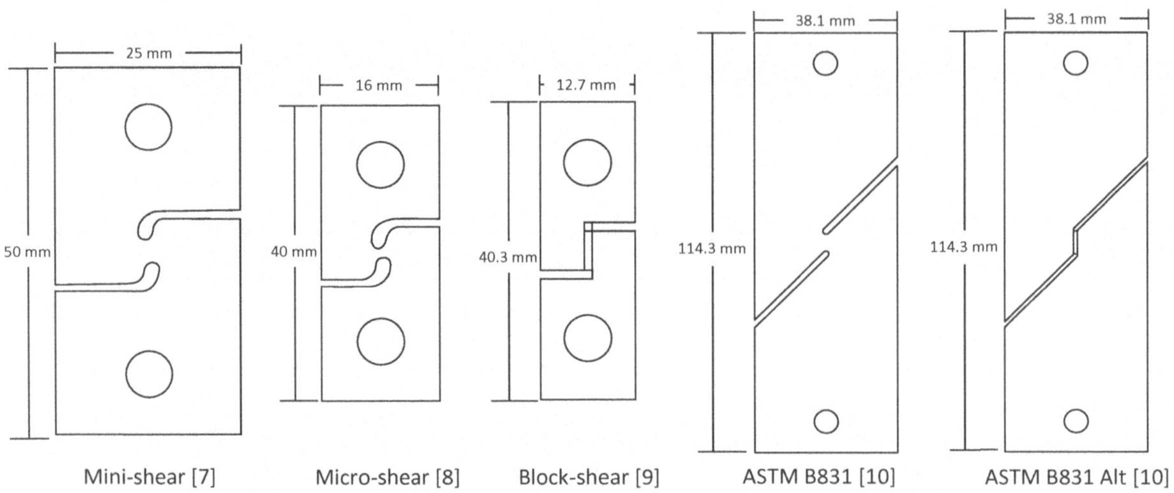

Fig. 11.1 Geometry of shear samples tested

11.2 Sample Preparation and Testing

The bulk adhesive samples were machined from 3.175 mm (1/8″) thick cast plaques of two-part toughened structural epoxy (3M 7333 Impact Resistant Structural Adhesive, 3M Canada Company, London, Canada). These plaques were created by pooling adhesive on a 304.8 mm × 304.8 mm (12″ × 12″) plate of tempered glass, taking care to minimize entrapped air. Spacers (3.175 mm thick) were then placed on all four sides of the pool of adhesive and a second plate of glass was placed on top of the entire assembly. The glass plates were clamped together and placed in a convection oven (ED-53, Binder, Tuttlingen, Germany) to cure for 30 min after reaching 80 °C. After curing, the plaques were CNC machined to the required geometry for each test sample. Note that for the block-shear and ASTM B831 Alt samples, the gauge area was further machined to a thickness of 1.4 mm and 1 mm respectively, while the remaining samples utilized the full 3.175 mm (1/8″) thickness.

After machining, the gauge length of each sample was measured using an opto-digital microscope (VHX 5000, Keyence, Osaka, Japan), which could resolve the dimensions to an accuracy of 50 μm. The gauge thicknesses of each mini-shear, micro-shear, and ASTM B831 sample were measured with a digital micrometer. The depth of material removed from gauge area of the block-shear and ASTM B831 Alt samples was measured using the opto-digital microscope. These measurements were then subtracted from the sample thickness to calculate the thickness of the gauge area.

Each test was imaged using a single DSLR camera at a resolution of 1920 × 1080 pixels with a 105 mm f/2.8 macro lens and 2X teleconverter. A monochrome speckle pattern was applied to the gauge area to allow 2D digital image correlation (DIC) to be used to measure shear strain. The DIC analysis was carried out at a frame rate of 5 fps with the resolution in the area of interest being 0.006 mm/pixel. A DIC subset size of 41 pixels, a step size (*i.e.*, the distance between subset centers during analysis) of 1, and a strain filter of 15 pixels were used for each analysis. The shear strain for each sample was reported from the central location of the gauge area as described in Rahmaan et al. [8].

Samples were tested at a deformation rate of 0.0254 mm/s (0.001 in./s) using a custom-made hydraulic load frame equipped with a 2224 N (500 lb) load cell. Force data was recorded at a frequency of 120 Hz. The load cell and video data were synchronized using an LED, which was illuminated at the start of data acquisition. After testing, the nominal shear stress was calculated by dividing the measured force by the gauge length and thickness measured for each sample.

11.3 Results

The stress-strain responses for the samples (Fig. 11.2) were consistent in shape with peak stresses ranging from 23 to 28 MPa. In contrast, there was a substantial amount of scatter in the strain-to-failure measurements between the samples (0.15–0.32 mm/mm) and in some cases, within each sample response (*e.g.*, 0.13–0.24 mm/mm for the mini-shear sample). The micro-shear sample and ASTM B831 Alt samples exhibited the highest average strain to failure of 0.32 mm/mm and 0.28 mm/mm, respectively.

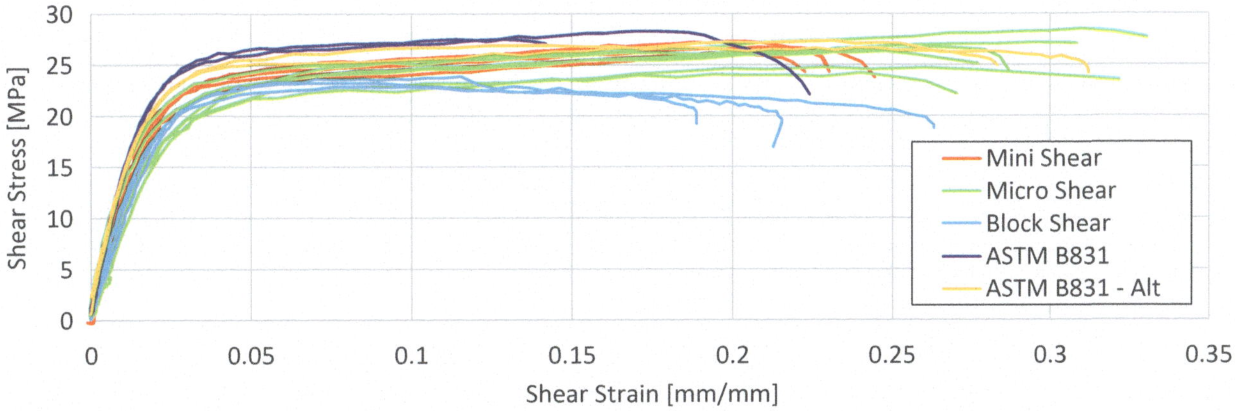

Fig. 11.2 Shear stress-strain response (5 geometries, 3 repeats)

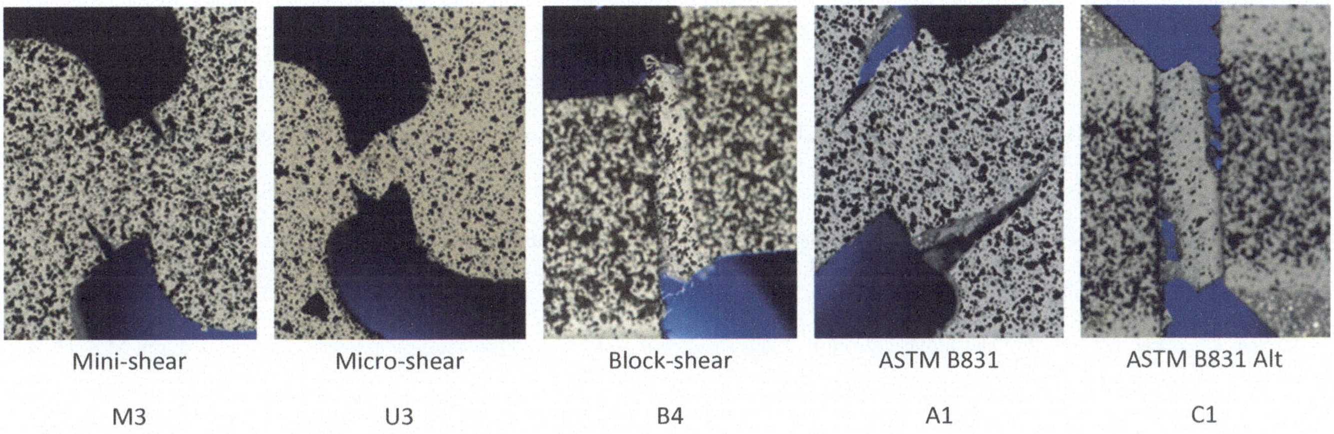

Fig. 11.3 Fracture location of each test geometry

The failure initiation of the mini-shear (Fig. 11.3a) and ASTM B831 (Fig. 11.3d) samples originated at locations of high tensile (first principal) strain, leading to fractures transverse to the direction of loading. This strong dependence on tensile strength is a common trait seen when testing polymeric materials under shear loading and has been demonstrated by several other studies using different test sample geometries [2, 4–6]. The failure of the micro-shear samples (Fig. 11.3b) initiated in a similar location as the mini-shear samples. However, due to the shorter gauge length, the failure initiation location was located closer to the gauge area, and tended to propagate through the gauge area rather than transverse to the direction of loading. Failure of the block-shear (Fig. 11.3c) and ASTM B831 Alt (Fig. 11.3e) samples initiated at the root of the stepped section of the sample, likely due to high stress concentrations present from the sharp corners introduced from machining. This was particularly pronounced in the block-shear sample, explaining the reduction in shear stress following yield.

It is worth noting that the lateral offset between the two cut out lobes of the mini-shear sample geometry was originally selected to ensure the center of the gauge was as close to pure shear as possible prior to failure [7]. The sample was originally designed for use with metals and optimized using a Johnson-Cook based material finite element model, which may not be appropriate for polymeric materials. The developers noted that the eccentricity was introduced to induce compressive stresses in the gauge area of the sample, which were necessary to counteract the high hydrostatic stresses that developed at the failure initiation location demonstrated in the current study. The authors concluded that higher eccentricity was necessary to ensure pure shear stress for materials with lower Young's modulus and that the geometry should be optimized for each material. Since the micro-shear sample was a scaled version of the mini-shear sample, this was also the case for that geometry.

11.4 Conclusions

Measurement of adhesive shear response is a critical input to accurately predict the response of bonded joints in finite element models [11]. The five sample geometries tested in this study all produced consistent stress-strain responses with similar shape, and average peak stress values within 5 MPa. The maximum strain measured for each sample exhibited large variability due to fracture initiating at locations of high tensile strain with the mini-shear, micro-shear, and ASTM B831 samples, and due to stress concentrations at sharp notches in the block-shear and ASTM B831 Alt samples. Of the samples tested, the micro-shear sample provided the largest average strain measurement prior to fracture initiation and thus the most complete stress-strain response for finite element modeling.

References

1. G'Sell, C., Boni, S., Shrivastava, S.: Application of the plane simple shear test for determination of the plastic behaviour of solid polymers at large strains. J. Mater. Sci. **18**(3), 903–918 (1983)
2. Morin, D., Haugou, G., Bennani, B., Lauro, F.: Identification of a new failure criterion for toughened epoxy adhesive. Eng. Fract. Mech. **77**(17), 3481–3500 (2010)
3. Farrokh, B., Khan, A.S.: A strain rate dependent yield criterion for isotropic polymers: low to high rates of loading. Eur. J. Mech. A Solids. **29**(2), 274–282 (2010)
4. García, J.A., Chiminelli, A., García, B., Lizaranzu, M., Jiménez, M.A.: Characterization and material model definition of toughened adhesives for finite element analysis. Int. J. Adhes. Adhes. **31**(4), 182–192 (2011)
5. Liu, K., Piggott, M.R.: Fracture failure processes in polymers. I: mechanical tests and results. Polym. Eng. Sci. **38**(1), 60–68 (1998)
6. Liu, K., Piggott, M.R.: Fracture failure processes in polymers. II: fractographic evidence. Polym. Eng. Sci. **38**(1), 69–78 (1998)
7. Peirs, J., Verleysen, P., Degrieck, J.: Novel technique for static and dynamic shear testing of Ti6Al4V sheet. Exp. Mech. **52**(7), 729–741 (2012)
8. Rahmaan, T., Butcher, C., Abedini, A., Worswick, M.: Effect of strain rate on shear properties and fracture characteristics of DP600 and AA5182-O sheet metal alloys. In: 11th International Conference on the Mechanical and Physical Behaviour of Materials under Dynamic Loading, vol. 94, p. 1033 (2015)
9. Gardner, K.: Experimental techniques for shear testing of thin sheet metals and compression testing at intermediate strain rates. M.S. Thesis, The Ohio State University (2013)
10. ASTM B831, B831-10 Standard Test Method for Shear Testing of Thin Aluminum Alloy Products, ASTM International (2014)
11. Kolling, S., Haufe, A., Feucht, M., Du Bois, P. A.: A constitutive formulation for polymers subjected to high strain rates. In: 9th international LS-Dyna Users Conference, pp. 15–55. Detroit (2006)

Brock Watson is a Ph.D. student at the University of Waterloo researching adhesive bonding of high-strength steels for structural crashworthiness.

Chapter 12
A Design of Experiments Approach for Determining Sensitivities of Forming Limit Analyses to Experimental Parameters

Mark A. Iadicola and Dilip K. Banerjee

Abstract Forming limit testing is widely used to characterize materials for use in forming operations. Subset-based Digital Image Correlation (DIC) has become a common method used in research laboratories to measure the forming limit strain fields, but no consistent method exists for forming limit determination from these data. In this work, we apply a design of experiments approach along with Finite Element Analyses to exercise potential experimental measurement requirements numerically (especially varying expected image resolutions and frame rates) to quantifying the sensitivities of various proposed forming limit analysis methods to changes in experimental setup and DIC processing parameters. The main goal is to determine the appropriate hardware to produce benchmark experimental data, and possibly highlight which forming limit analysis methods may be either overly sensitive, or preferably insensitive, to subjective processing decisions.

Keywords Forming limit · Sheet metal · DIC · Design of experiments · FEA

12.1 Background

Forming limit testing is widely used to characterize materials for use in forming operations. The result of the test is a material's forming limit curve (FLC) that is drawn in strain space. The FLC denotes the boundary between a safe and unsafe level of plastic strain achieved through nominally linear strain-ratio paths. During the test, a sheet is locked in an axi-symmetric binder and deformed until necking or failure [1–3] using either a hemispherical dome (i.e., Nakajima type) or cylindrical punch with a radius edge (i.e., Marciniak type).

12.2 Motivation

There has been a concerted effort to improve the repeatability and reliability of the forming limit test results through clearly defined procedures in conjunction with the use of digital image correlation (DIC) or similar strain field measurements [3]. Obviously, a clearly defined FLC analysis method for the measured strain field information from each test is required. Although there is a defined analysis method described in ISO 12400-2 [3], it is not widely accepted. This results in a standard test procedure that is followed, but often with a non-standard FLC analysis method. Many FLC analysis methods have been proposed and demonstrated on specific material classes using specific DIC hardware, but sometimes these procedures cannot be used on other material classes or using other common hardware. Generally, the analysis methods fall into two main categories; those focused on spatial analysis of the strain field at a specific time in the strain history (e.g., [3, 4]), or those focused on a temporal analysis of the strain at one or two specific locations on the sample (e.g., [5]).

Not surprisingly, each FLC analysis method has advantages and disadvantages, but they also each have unintended assumptions and requirements regarding the DIC strain measurements to be analyzed. DIC can create a data set that could be used for either a spatial or temporal analysis, but the effect of DIC system hardware and processing parameters are often ignored, even though they can have substantive effects on the DIC strain fields. The choice of the appropriate FLC analysis method for a material, or intended use, is not obvious. One possible path forward is to develop a benchmark set of forming limit tests with DIC image acquisition and associated calibrations that researchers could analyze with their preferred DIC software, various DIC processing parameters, and proposed FLC analysis method. This is somewhat similar

M. A. Iadicola (✉) · D. K. Banerjee
National Institute of Standards and Technology, Gaithersburg, MD, USA
e-mail: mark.iadicola@nist.gov

M. Grady et al. (eds.), *Mechanics of Biological Systems & Micro-and Nanomechanics, Volume 4*, Conference Proceedings of the Society for Experimental Mechanics Series, https://doi.org/10.1007/978-3-319-95062-4_12

to the DIC Challenge [6] concept, but more focused on a test method and its interpretation, and less on the DIC measurements themselves. Setting up an experiment to create such a benchmark set is not trivial. Many methods seemingly require either very fine spatial resolution in their images or very fine temporal sampling. The initial range of DIC setup properties being considered includes: camera resolution, lenses, lighting, stereo angle, DIC pattern, exposure time, imaging rate, frame timing uncertainty, DIC system calibration, and air currents.

In addition to the DIC hardware (and setup), the DIC processing parameters can have a large influence on the position and strain measurements. Various artifacts resulting from processing parameters exist. For example, [7, 8] demonstrated that the displacement errors vary with subset size and pattern quality, and [7] also showed that the subset size can spatially smooth the measured 3D shape of a test piece. It is not clear how individual aspects of hardware and processing will affect the FLC determination. The initial DIC processing parameters to be considered include: subset size, subset overlap (step size), subset shape function, interpolant, strain calculation method, strain filtering. Some are clearly selectable (e.g., subset size), but some artifacts that need to be considered are the result of a combination of the DIC hardware and processing (e.g., measurement in-plane uncertainty). There are both obvious and subtle interactions between many of these variables, and although some of these interactions are deterministic others can be stochastic.

12.3 Proposed Solution

Here we propose a numerical approach using the Design of Experiments (DOE) technique to determine the sensitivity of the FLC result using a combination of FEA, DIC setup and processing parameters, and various FLC analysis methods. The DOE approach should allow the study of the influence of input parameters on the resulting FLC, identification of possible input parameter interactions, efficient evaluation of a multi-parameter design space and estimate of an optimum design point, and develop an approximate model of the sensitivities within the design space. The DOE results can be viewed through main effects plots, interaction effects plots, correlation graphs, and optimum design point determinations. These results will be used to select the hardware for the benchmark testing, and should illuminate the strengths and weaknesses of the various FLC analysis methods.

Since each analysis method is known to result in a different FLC, they cannot be checked against a known reference result. However, they can be checked for their individual sensitivity to various parameters, and they can be checked against FLC determined based on an ideal set of strain data with no artifacts of DIC hardware or processing parameters. To that end, we assume that a Finite Element Analysis (FEA) (e.g., Fig. 12.1a shows a mesh for a uniaxial test piece) can produce a set of data at a refinement beyond any desired measurement precision (e.g., camera resolution and/or frame rate) and without error (e.g., noise free) to be analyzed for FLC to serve as the reference value that should be predicted after the measurement artifacts have been added. These artifacts will be abstracted to be easily implemented through processing of the FEA data. For example, the subset size will be used to average the position data over a patch similar to the subset size, and noise will be added based on experience for that subset size in conjunction with a typical DIC pattern (see [7, 8]). More advanced methods could be applied similar to [9], but the added complication would add substantial computation time for what might be minimal additional understanding. This is left to a possible future extension of the work. In addition to computational time, there are also storage issues (write out and read in) that affect the overall computational time. Although the candidate FLC methods require a wide range of data from the test, there is substantial overlap regarding the data needed. Additionally, the data from an actual test using DIC would only be surface displacement information. These two factors allow the determination of FLC from a much smaller data set than the entire FEA model. Figure 12.1a shows this surface area set in grey on the initial mesh that is centered on the area that begins to neck in the model. The DOE will operate on the results calculated at the nodes within this surface set of data. Figure 12.1b is a schematic of the data used in the FLC determination. Generally, the spatial methods use profile samples of the DIC data (black dashed lines in Fig. 12.1b) on a specific image (typically one or two before failure), whereas the temporal methods will track some behavior at a specific point through the time of the test (color dashed lines with arrows in Fig. 12.1b). The grey area shown in Fig. 12.1 is required to extend beyond the dashed lines, due to the size of the DIC subset that may be used to measure the displacements and calculate strains.

The numerical process is shown in Fig. 12.2 as a flowchart. In order to determine a FLC reference value and permit a consistent basis for comparison of FLC results, the FEA is only run one time for a given strain ratio (e.g., uniaxial in Fig. 12.1a) at a very refined spatial and temporal resolution with the results written to disk only for the critical portion of the test piece (grey area in Fig. 12.1a). Even when writing only the limited data set (grey area in Fig. 12.1a over time), initial work demonstrated that the required data precision and refinement of the FEA model requires that the data be written in binary format for reasonable scalability of the problem. Before the DOE process cycle begins, a DOE parameter set matrix (lower left in Fig. 12.2) is developed to efficiently sample the multi-parameter design space. The number of combinations

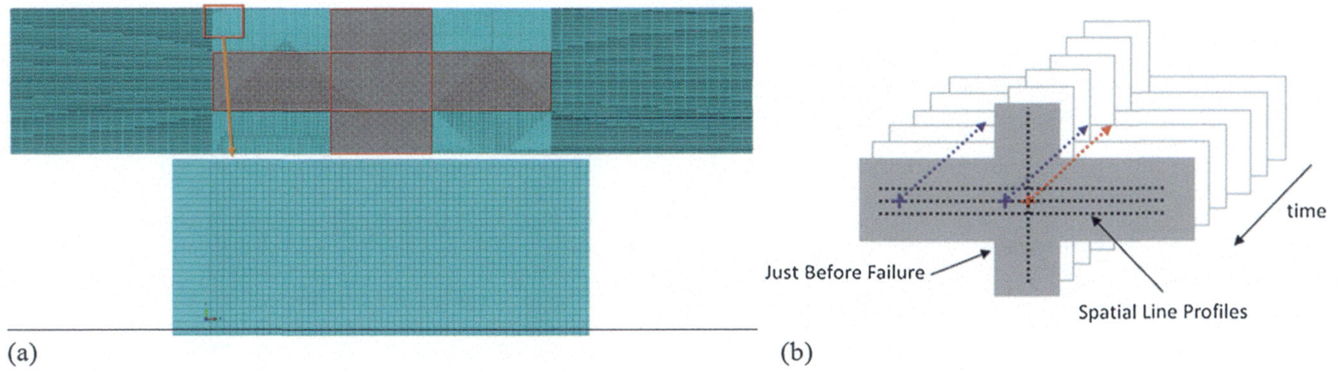

Fig. 12.1 (**a**) FEA mesh of a uniaxial specimen with sampling on the surface (grey cross) to be used in DOE. (**b**) Schematic of data used in DOE for the various FLC determinations: spatial (black dashed lines) and temporal (color dashed lines)

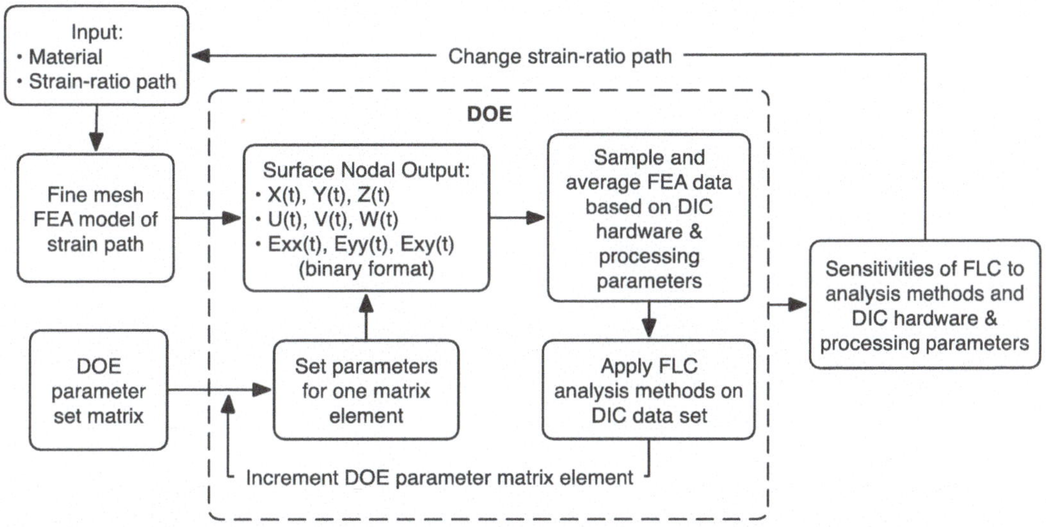

Fig. 12.2 Flow chart of the Design of Experiments (DOE, dashed rectangle) including the Finite Element Analysis (FEA), Forming Limit Curve (FLC) determination method, and the Digital Image Correlation (DIC) parameters

of input factors is user selectable, but must be greater than the number of parameters to be varied. For each run, the DOE process cycle takes one of these DOE parameter sets from the matrix and uses it to select the data to be processed from the FEA output. These data are then processed through Matlab[1] scripts of the abstractions for the DIC hardware and processing parameters (e.g., camera resolution and subset size averaging) based on the selected DOE parameter set. After the DIC processing step, the data are passed to the routines for all the FLC analysis methods under consideration resulting in a single FLC value for each. Then the next DOE matrix parameter set is selected and the process is repeated. In addition to the FLC values determined for each DOE matrix set, the resulting sensitivities and parameter interactions can be assessed (right hand side of Fig. 12.2) after completing the entire DOE matrix. Those results should determine the appropriate hardware to produce benchmark experimental data, and possibly highlight which FLC analysis methods may be either overly sensitive, or preferably insensitive, to subjective processing decisions.

References

1. ASTM: Standard Test Method for Determining Forming Limit Curves, E2218-16, ASTM International (2016)
2. ISO: Metallic materials-sheet and strip-determination of forming-limit curves-Part 1: Measurement and application of forming- limit diagrams in the press shop, ISO12004-1:2008 (2008)

[1]Certain commercial software is identified to describe a procedure or concept adequately. Such identification is not intended to imply recommendation, endorsement, or implication by NIST that the software or materials are necessarily the best available for the purpose.

3. ISO: Metallic materials-sheet and strip-determination of forming-limit curves-Part 2: Determination of forming-limit curves in the laboratory, ISO12004-2:2008(E) (2008)
4. Situ, Q., et al.: A suitable criterion for precise determination of incipient necking in sheet materials. In: 10th International Conference on Aluminium Alloys (ICAA- 10), Vancouver, CANADA, 09-13 Jul 2006. Materials Science Forum, vol. 519–521, pp. 111–116 (2006).
5. Sriram, S., et al.: Comparison of forming limit curves for advanced high strength steels using different techniques. SAE Int. J. Mater. Manuf. **2**(1), 472–481 (2009)
6. Reu, P.L., et al.: DIC challenge: developing images and guidelines for evaluating accuracy and resolution of 2D analyses. Exp. Mech. (2017). Doi:https://doi.org/10.1007/s11340-017-0349-0
7. Iadicola, M.A.: Augmented use of standard mechanical testing measurements for sheet metal forming: digital image correlation for localized necking. In: Yoon J.W., et al. (eds.) NUMISHEET 2014 Part B General Papers. AIP Proceedings 1567, American Institute of Physics, Melville, NY, pp. 614–619 (2014)
8. Iadicola, M.A., Creuziger, A.A.: Uncertainties of digital image correlation near strain localizations. In: Proceedings of the SEM Annual Meeting (2014)
9. Balcaen, R., et al.: Stereo-DIC calibration and speckle image generator based on FE formulations. Exp. Mech. **57**(5), 703–718 (2017). https://doi.org/10.1007/s11340-017-0259-1

Mark A. Iadicola is a staff scientist at NIST for over a decade, and is currently a member of the Materials Science and Engineering Division, NIST Center for Automotive Lightweighting.

Chapter 13
Post Processed Foundry MEMS Actuators for Large Deflection Optical Scanning

La Vern Starman, D. Torres, H. J. Hall, J. P. Walton, and R. A. Lake

Abstract In this research effort, we developed MEMS micromirrors which are suitable for large angle deflections for optical scanning applications using the PolyMUMPs™ foundry fabrication process. This foundry process uses polysilicon and gold as the structural layers to form the bimorph beam structures which make up our actuation assembly. From both modeling and experimental testing, current micromirror designs fabricated in the PolyMUMPs™ process do not meet the deflection requirements to enable large angle scanning. As a result, we developed several post processing deposition techniques, using the PolyMUMPs™ structural layers as the baseline structure, to enable the necessary upward deflections to permit large angle scanning. In this research, we design, model, post fabricate, and test high out-of-plane MEMS actuators intended for integration with SOI micromirror arrays to enable the broadband, high fill-factor scanning applications. These arrays are designed to scan in multiple directions due to the segmented actuation design. The upward deflection is accomplished through material selection and design control (i.e., structure length, material thickness, material coefficient of thermal expansion (CTE), deposition temperature, and material layer composition) of bimorph structures. Following the post processed fabrication and sacrificial release; the initial deflection profiles are measured and compared against COMSOL™ models.

Keywords MEMS · Beamsteering · Micromirror · Post processed · Finite element modeling (FEM)

13.1 Introduction

Micro-Electro-Mechanical System (MEMS) micromirrors have been in development since the 1990s; however, large angle beamsteering using micromirrors is a fairly recent application space that only a few researchers have delved into [1–7]. Micromirrors are the technology of choice over liquid crystals and semiconductor based methods through careful design, offer several advantages including minimal power consumption, a wide optical bandwidth, and a dynamic response suitable for most applications. However, the design and fabrication of large, out-of-plane micromirror structures which also includes a high fill factor (>90%) are in the early stages of development. Literature reveals a few large angle beamsteering approaches but none exhibit a high fill-factor with most micromirror designs set to steer <±20° [2–7]. To address the current limitations in large angle beamsteering, this work focuses on achieving a device that exhibits both tip and tilt beamsteering angles approaching ±45°, a fill factor >90% and a switching speed around 1 ms. To create the large, upward deflections needed for this large angle beamsteering effort, bimorph structures which leverage the thermal stress created from the coefficient of thermal expansion (CTE) mismatch of thin film layers is employed. To accomplish the large deflections, material selection and design control (i.e. structure length, material thickness, material coefficient of thermal expansion (CTE), deposition temperature, and material layer composition) of the bimorph based structures is critical.

In general, MEMS designers are limited to the materials and fabrication processes available in their fabrication facility or that of a commercial foundry. In this work, the PolyMUMPs™ foundry fabrication process is leveraged which uses only polysilicon and gold as the structural layers [8]. The limitations of this fabrication process alone are insufficient to meet device design goals needed for large angle steering, namely initial out-of-plane deflection. Since our large angle beamsteering approach intends to take advantage of the inherent residual stresses and high CTE levels in bimorph devices, our designs require post processing to increase the initial upward deflection levels to enable large angle beamsteering. Thus,

L. V. Starman · D. Torres (✉) · H. J. Hall · J. P. Walton
Sensors Directorate, Air Force Research Laboratory, Wright-Patterson AFB, OH, USA
e-mail: david.torres-reyes.1@us.af.mil

R. A. Lake
Air Force Institute of Technology, Wright-Patterson AFB, OH, USA

M. Grady et al. (eds.), *Mechanics of Biological Systems & Micro-and Nanomechanics, Volume 4*, Conference Proceedings of the Society for Experimental Mechanics Series, https://doi.org/10.1007/978-3-319-95062-4_13

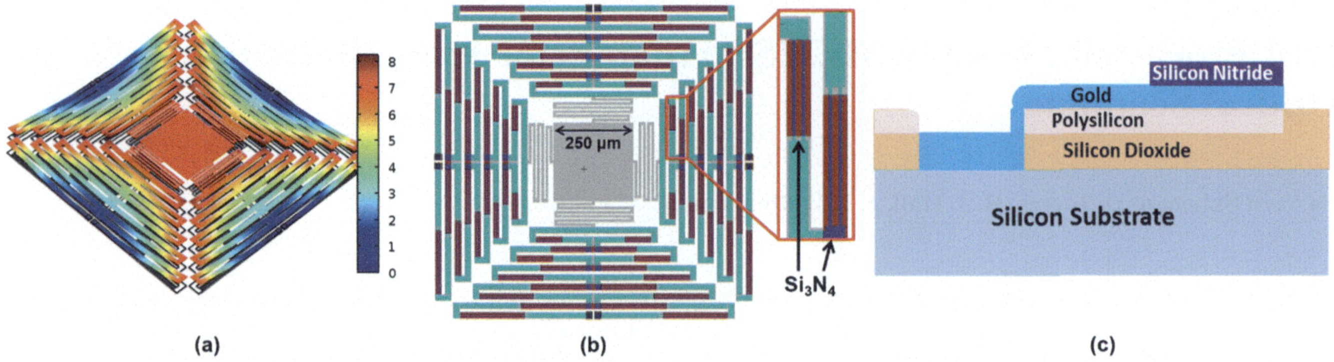

Fig. 13.1 (**a**) FEM simulation demonstrating the deformation of the design without post-processing. (**b**) Illustrates our ~1 mm^2 L-Edit rendition of the PolyMUMPs™ electrothermal actuator with the additional Si_3N_4 layer shown in purple; this image also shows the locations where the post processing Si_3N_4 layer will be used to enable the upward deflection. An abbreviated cross-sectional view of the PolyMUMPs™ fabrication process with the post processed Si_3N_4 layer before the release of the structure is shown in (**c**)

we integrated a 1 μm thick high stressed plasma enhanced chemical vapor deposition (PECVD) silicon nitride (Si_3N_4) layer. Our initial design was accomplished using commercial finite element modeling (FEM) software (COMSOL™ Multiphysics v 5.2a), utilizing material properties provided by MEMSCAP® [9]. The residual stress value from MEMSCAP® for the polysilicon layer is ~−7 MPa compressive stress while the gold metal layer has a tensile stress level of 24 MPa. Baseline (zero applied voltage) steady-state deformation simulations of designs without post-processing were accomplished for initial model validation. Simulated results indicated no initial upward deflection of the platform assembly which is shown in Fig. 13.1a. Figure 13.1b illustrates our ~1 mm^2 L-Edit rendition of the PolyMUMPs™ electrothermal actuator with the additional Si_3N_4 layer shown in purple; this image also shows the ideal locations where the post processing Si_3N_4 layer will be used to enable the upward out-of-plane deflection. Figure 13.1c provides a partial cross-section template of the PolyMUMPs™ fabrication process with the addition of the post processed Si_3N_4 layer.

13.2 Design Geometry and Simulated Results

From the initial simulated results, it was observed that the PolyMUMPs™ fabrication process alone will not meet the out-of-plane deflection goals, thus post processing is required. Figure 13.1b provides an image of the L-Edit design for the overall 10-beam electrothermal actuation assembly as modeled in COMSOL™ with the addition of the Si_3N_4 layer shown in purple. The actual dimension of this actuation assembly is ~1 mm^2. Figure 13.2a shows the peak deflection of the simulated COMSOL™ model using the MEMSCAP® stress values obtained for the polysilicon and gold structural layers with the addition of the ~−1.5 GPa layer of Si_3N_4. The simulation results illustrate an initial upward out-of-plane deflection of 279.8 μm. Figure 13.2b shows the overall stress levels across the actuation assembly. Based on the stress map from COMSOL™, shown in Fig. 13.2b, one can observe there are no high stress concentrations along the beam structures or across the actuation assembly which helps mitigate localized fracture points during actuation. From the inset in Fig. 13.2b, our peak stress is ~4 GPa which is significantly less than the published experimental fracture stress level in polysilicon of ~5.4–6.0 GPa [10].

13.3 Experimental Results

The L-Edit design as shown in Fig. 13.1a was fabricated in the PolyMUMPs™ foundry process. Once the die were received, a 1 μm thick, highly stressed Si_3N_4 layer was deposited and precisely patterned as illustrated in Fig. 13.1b to validate the significant upward out-of-plane deflections as shown if Fig. 13.2a. This high deflection was verified through the post processing deposition of the Si_3N_4 layer and is shown in Fig. 13.3a. Figure 13.3b shows a 3D image obtained by an Olympus LEXT laser microscope of the MEMS actuation structure, where the released structure shows a 273.39 μm of upward out-of-plane deflection. The Olympus system is used to measure the high deflection profiles for our devices and to measure the tilt/tip angles of the actuation platform.

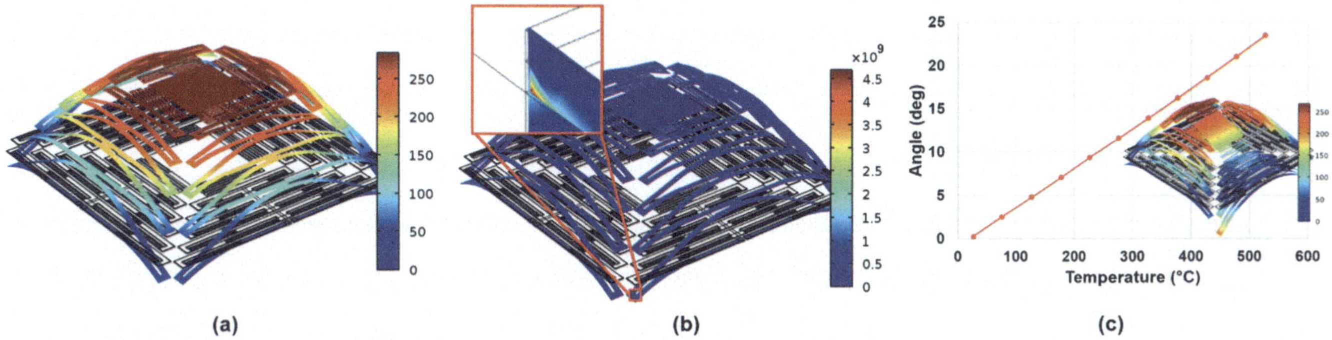

Fig. 13.2 FEM simulation results of the release structure with the Si_3N_4 layer: (**a**) out-of-plane initial deflection of the platform: 279.8 μm, and (**b**) overall stress levels across the actuation assembly, and (**c**) simulated steady-state actuation of the device by artificial uniform temperature application of single actuator (not electrical joule heating)—an angle of 23.5° is achieved at 526 °C

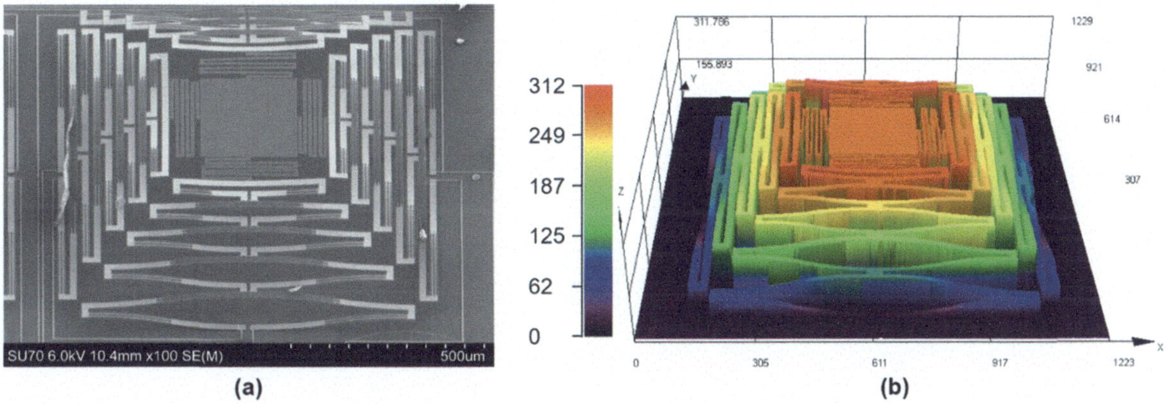

Fig. 13.3 (**a**) An SEM image of the released MEMS structure with the added Si_3N_4 illustrating the large, upward, out-of-plane deflection. (**b**) Illustrates a 3D image of the released structure obtained from an Olympus LEXT laser microscope where the elevation of the platform was measured: 273.39 μm

13.4 Conclusions

The realization of a large angle beamsteering actuation assembly was presented which showed the necessary upward out-of-plane deflection can be achieved through the use of a high stressed Si_3N_4 material layer selectively patterned and etched when placed at key locations on top of the actuation beam assembly. COMSOL™ was used to simulate the pre and post processed addition of the Si_3N_4 layer and was validated through experimental test on PolyMUMPs™ fabricated structures. The pre Si_3N_4 deposition results in no out-of-plane deflection of the platform assembly while the post processed Si_3N_4 layer deposition measures an out-of-plane deflection of 273.39 μm. From these experimental results, COMSOL™ models do provide realizable estimations of the initial upward deflections based solely on the material properties and layer compositions of the beams which make up the actuation assembly. Future research will focus on the performance of the actuation assembly to determine peak tip/tilt and piston motion and determine the switching frequency and perform a thermal evaluation of the actuation assembly.

Acknowledgements Support for this research is greatly appreciated with funding through the Air Force Research Laboratory, Aerospace Component and Subsystem Division Titled: "Design, Model, and Fabrication of a 5 × 5 Large Tip, Tilt, and Piston MEMS Micromirror Array".

References

1. Yoo, B., Megens, M., Chan, T., Sun, T., Yang, W., Chang-Hasnain, C., Horsley, D., Wu, M.: Optical phased array using high contrast gratings for two dimensional beamforming and beamsteering. Opt. Express. **21**(10), 12238–12248 (2013)

2. Samuelson, S., Xie, H.: A large piston displacement MEMS mirror with electrothermal ladder actuator arrays for ultra-low tilt applications. J. Microelectromech. Syst. **23**(1), 39–49 (2014)
3. Sandner, T., Grasshoff, T., Schenk, H., Kenda, A.: Out-of-plane translatory MEMS actuator with extraordinary large stroke for optical path length modulation in miniaturized FTIR spectrometers. In: Sensor+Test Conferences 2011 – IRS^2 Proceedings, pp. 151–156 (2011)
4. Todd, S., Jain, A., Qu, H., Xie, H.: A multi-degree-of-freedom micromirror utilizing inverted-series-connected bimorph actuators. J. Opt. A Pure Appl. Opt. **8**, 352–359 (2006)
5. Wu, L., Dooley, S., Watson, E., Mcmanamon, P.: A tip-tilt-piston micromirror array for optical phased array applications. J. Microelectromech. Syst. **19**(6), 1450–1461 (2010)
6. Jia, K., Pal, S., Xie, H.: An electrothermal tip-tilt-piston micromirror based on folded dual S-shaped bimorphs. J. Microelectromech. Syst. **18**(5), 1004–1015 (2009)
7. Lim, S., Choi, J., Horowitz, R., Majumdar, A.: Design and fabrication of a novel bimorph micro-opto-mechanical sensor. J. Microelectromech. Syst. **14**(4), 683–690 (2005)
8. Carter, J., Cowen, A., Hardy, B., Mahadevan, R., Stonefield, M., Wilcenski, S.: PolyMUMPs™ design handbook. (Revision 13). [On-Line]. http://memscap.com/__data/assets/pdf_file/0019/1729/PolyMUMPs-DR-13-0.pdf. Accessed 26 Jan 2018
9. Cowan, A.: PolyMUMPs. http://www.memscap.com/products/mumps/polymumps. Accessed 26 Jan 2018
10. Brueckner, J., Dehe, A., Auerswald, E., Dudek, R., Vogel, D., Michel, B., Rzepka, S.: Investigating fracture strength of polysilicon membranes using microscopic loading tests and numerical simulation. Microsyst. Technol. **22**(3), 569–575 (2016)

La Vern Starman is the lead engineer in the optical beamsteering group at AFRL in the Highly Integrated Microsystems Branch at Wright-Patterson AFB. He received his B.S. from the University of Nebraska, his M.S. from Wright State University and his Ph.D. from AFIT all in EE.

Chapter 14
Torsional Structures to Enable Large Angle Deflections

La Vern Starman, D. Torres, and J. P. Walton

Abstract MEMS micromirrors for large angle beamsteering are needed for numerous broadband steering and imaging applications. However, current scanning micromirrors generally exhibit scanning angles (<20°) and typically in only one direction. In addition, the optical fill-factor is generally <50% which leads to significant signal loss and possible heating of the underlying mechanical structures. In this research, we design, model, fabricate, and test high; out-of-plane MEMS actuators utilizing torsional spring assemblies to enable the large angle (∼45°) tip-tilt of the actuation platform assembly. The micromirror exhibit tip, tilt, and piston motion due to the segmented actuation design approach used; thus, the torsional spring structural elements must be integrated into the design to enable the large tilt angles. The torsional structures must enable large angle bending while also being robust enough to support a micromirror structure during switching events. Through precise material selection, and design control (i.e., structure length, material thickness, Young's Modulus, and deposition temperature), these spring-based torsional structures can be used to enable the twisting and bending of the micromirror platform to enable the large angle scanning. Several different structural designs are assessed to determine the bending and torsional capabilities necessary for reliable and repeatable switching of the micromirrors.

Keywords MEMS · Beamsteering · Micromirror · Torsional springs

14.1 Introduction

Using Micro-Electro-Mechanical Systems (MEMS) as a technique for large angle beamsteering is a fairly recent application space with few researchers investing in this research focus area [1–7]. To enable the requested large beamsteering angles approaching ±45°, this research effort focuses on the design, fabrication, and development of the torsional spring attachments which exist between our coupled beam actuation assembly and the assembly platform. The coupled beam actuation assembly is composed of bimorph beam structures which capitalize on the materials inherent residual stress and the coefficient of thermal expansion (CTE). The residual stress and the materials CTE are essential to create the large, upward out-of-plane deflections required for this MEMS surface micromachined design. This large out-of-plane deflection enables the segmented actuators to individually pull down and rotate about the center axis to create the large tip/tilt angles on the platform assembly. Several design control metrics are necessary to achieve the appropriate torsional and bending tolerances to include, material selection, torsional spring segment lengths and widths, material thickness and the materials Young's Modulus.

In this research, both the PolyMUMPs™ foundry fabrication process [8] and an in-house fabrication process are used to create these large out-of-plane actuation assemblies which include a torsional spring attachment to the actuation platform as shown in Fig. 14.1. In Fig. 14.1, the left image illustrates the complete actuation assembly as fabricated in the PolyMUMPs™ fabrication process with the platform assembly identified. Figure 14.1b shows a close up view of the platform assembly and Fig. 14.1c shows the actual torsional spring structure.

The primary difference between the PolyMUMPs™ foundry fabrication and the in-house fabrication processes is centered on the materials used to create the torsional spring assembly. In the PolyMUMPs™ foundry fabrication process, a 1.5 μm thick polysilicon layer makes up the torsional spring assembly while a 2.0 μm thick silicon dioxide layer makes up the spring assembly in the in-house fabrication process. An advantage to our in-house fabrication effort is the torsional spring assembly could be made in a wide variety of materials to include silicon nitride, gold, aluminum, titanium, etc. to name a few. In essence, any fabrication friendly and fabrication compatible material can be used for these torsional spring attachment

L. V. Starman · D. Torres (✉) · J. P. Walton
Sensors Directorate, Air Force Research Laboratory, Wright-Patterson AFB, OH, USA
e-mail: david.torres-reyes.1@us.af.mil

M. Grady et al. (eds.), *Mechanics of Biological Systems & Micro-and Nanomechanics, Volume 4*, Conference Proceedings of the Society for Experimental Mechanics Series, https://doi.org/10.1007/978-3-319-95062-4_14

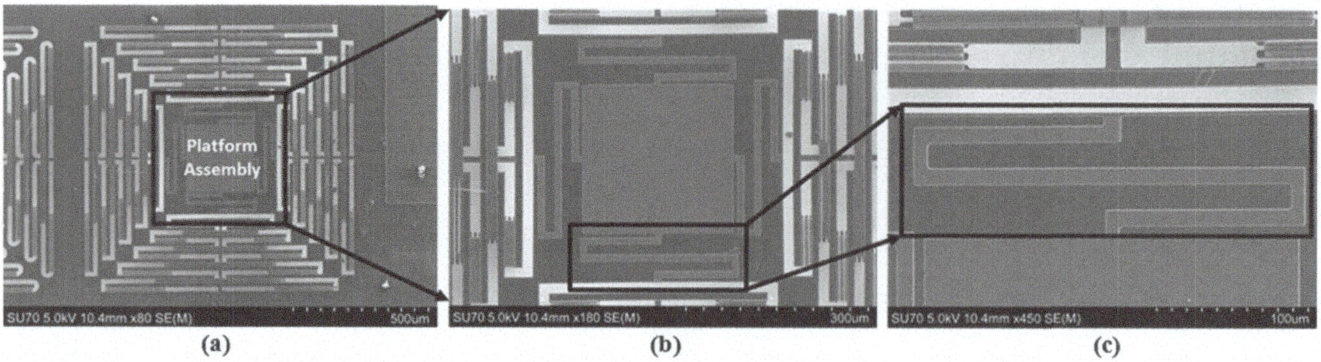

Fig. 14.1 (**a**) An SEM top view of the overall actuation assembly with platform/torsional spring assembly identified, (**b**) close up view of the platform assembly with torsional spring attachments, (**c**) close up view of a single torsional spring based attachment to permit tip/tilt motion

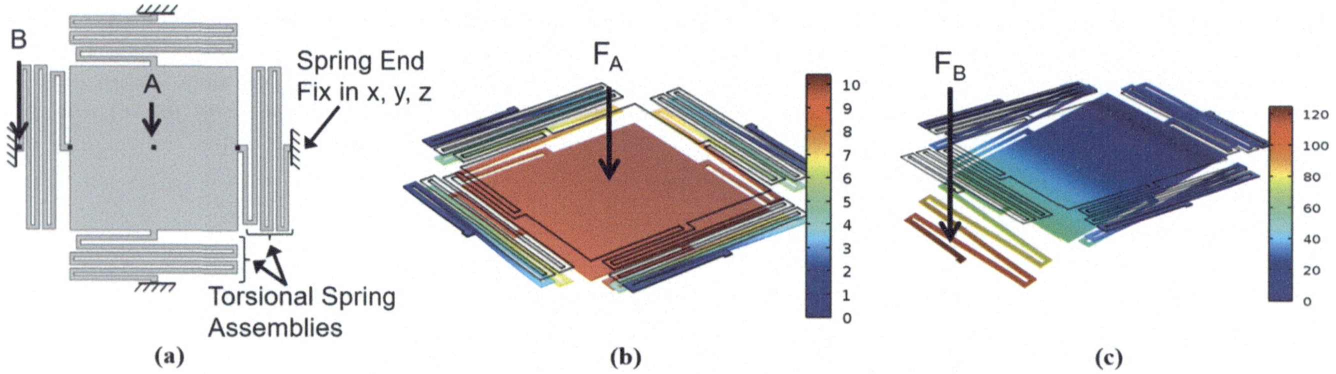

Fig. 14.2 (**a**) COMSOL™ image showing the torsional spring assemblies as attached to the platform. Points A and B represent the downward force locations to simulate the equivalent actuation forces to mimic the piston and tip/tilt actuation scenarios. (**b**) A force of 2 μN applied at Point A is used to observe the rigidity of the structure for the piston actuation, where the scale is in micrometer, and (**c**) Image with a force of 2 μN applied to Point B simulating a tip/tilt event, where the scale is in micrometer

structures; however, depending on the materials Young's modulus, the overall spring assembly may need to be slightly modified to adjust for the stiffness of the spring assembly (i.e., thinner material layer, length of spring beams) to enable proper bending and twisting.

14.2 COMSOL™ Modeling and Experiment Results

COMSOL™ finite element modeling (FEM) software was used to help determine the optimal torsional spring attachment which requires a relatively low tip/tilt spring constant but a fairly high piston spring constant. To satisfy the goals of this research, we developed and modeled several different torsional spring attachments which could possibly meet our project goals. Figure 14.2 shows the full design model of one torsional spring candidate and the associated images for piston and tip/tilt deflections for an applied force normal to the platform surface. To simplify the modeling, the physical actuators were removed and the attachment point normally connected to the actuation assembly is fixed in x,y,z as shown in Fig. 14.2a, where the fix point near the point B is removed when the force is applied at that location. Figure 14.2b illustrates the rigidity of the torsional springs for the piston motion when the force is applied at Point A. The main focus on the torsional spring attachment was to design a spring which would enable the large twisting of the actuation platform while also having the rigidity to prohibit the opposite actuator assembly during a tip/tilt event from deflecting downward toward the substrate. This simulation is shown in Fig. 14.2c where the force is applied at Point B, creating the twisting downward motion indicated by the extended spring on the left side of the image and the right side of the image remaining at approximately the original location. Any downward deflection of the opposite actuator assembly (right side shown in the image of Fig. 14.2c) will reduce the overall tilt angles that can be achieved as the physical rotating space will not be sufficient.

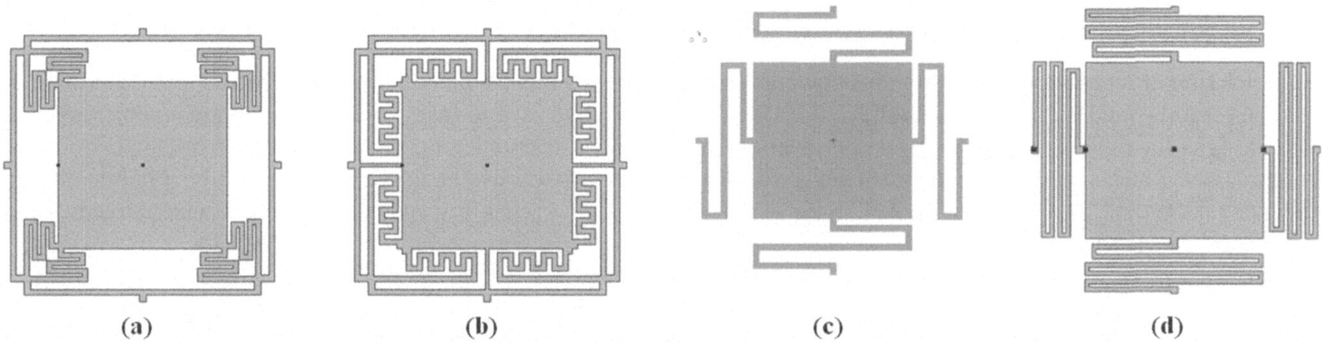

Fig. 14.3 Different torsional spring attachment configurations: (**a**) illustrates a torsional spring attachment which exhibits a moderately high piston and tip/tilt spring constants which will make tilt of the micromirror platform difficult, (**b**) illustrates the highest spring constants which make for a very ridged and stiff structure that nearly prohibits tilting events. (**c**) illustrates a lower spring constant which enables tilting but is not ridged enough such that it creates some of the negative downward motion of the opposite actuator assembly. Lastly, (**d**) provides a lower spring constant for a tip/tilt event and is rigid enough for piston actuation

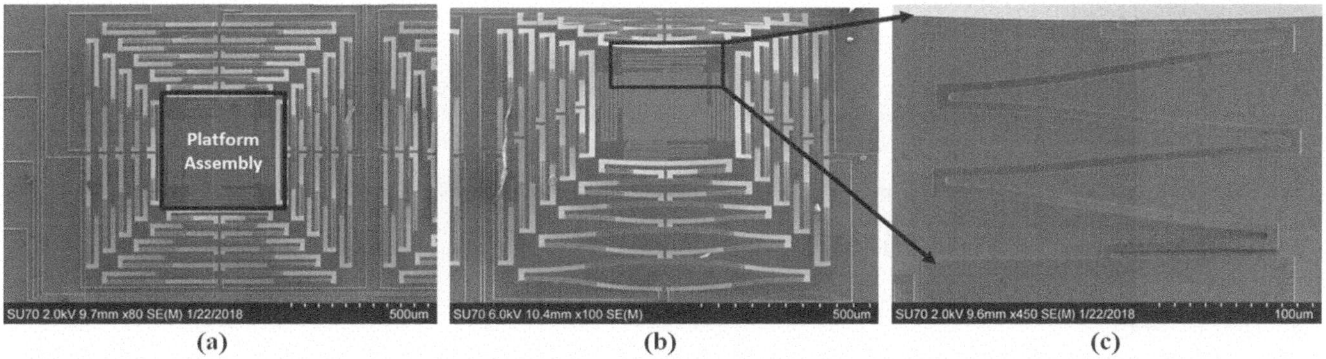

Fig. 14.4 (**a**) SEM image of the completed, unreleased PolyMUMPs™ micromirror actuation assembly with the platform assembly identified. (**b**) An SEM image of the same device in (**a**) following sacrificial oxide release. As can be seen, the platform assembly is deflected ~273 μm out-of-plane with the spring assembly boxed and referenced in (**c**) where the torsional spring is fully extended under an actuation condition

Figure 14.3 provides additional L-Edit torsional spring design attachments which can be used which exhibit varying degrees of torsional and twisting stiffness. The platform/spring assembly shown in Fig. 14.3a provides a moderately high spring constant for both the piston and tip/tilt motions, Fig. 14.3b provides a high spring constant which is ideal for rigidity and reliability of the platform assembly but also nearly mitigates any piston or tip/tilt motions when a 2 μN force is applied at Point A and B as shown in Fig. 14.2a. Figure 14.3c illustrates a lower spring constant which makes the platform motions resulting from forces applied to Points A and B greater but still results in the opposite actuator moving downward, reducing the peak tilt angle achievable. Figure 14.3d provide the best spring constants for this project as it allows for significant deflection of the platform resulting from either a force applied to Point A or B. However, the spring constants are high enough to maintain structural integrity following either actuation event.

The images shown in Fig. 14.4 show our complete micromirror design illustrating the torsional spring/platform assembly design as shown in Fig. 14.3d with the platform assembly boxed in Fig. 14.4a. Figure 14.4a provides a top image of the complete unreleased micromirror actuation design with the actuation assembly and platform assembly integrated to form one element of a micromirror actuation array. The structures shown in Fig. 14.4 were fabricated in the PolyMUMPs™ process such that the torsional spring attachment is ~1.5 μm thick with a width of 8 μm. Figure 14.4b illustrates the released micromirror actuation assembly which is deflected ~273 μm out-of-plane with the spring attachment boxed and reimaged in Fig. 14.4c when the spring is fully deflected. As can be observed in Fig. 14.4c, the spring is flexible enough to not fracture during full actuation conditions.

14.3 Conclusions

This realization of several torsional spring attachments were presented which is the critical linkage between the micromirror actuation assembly and the platform assembly. COMSOL™ models were used to assess the viability of the various torsional spring designs for rigidity and flexibility to perform piston motion as well as for tip/tilt motion. The fabricated structures were presented which clearly shows the torsional spring does not fracture when in its fully extended position. Future research will focus on the performance of the overall spring system to assess their reliability and functionality when operated by the micromirror actuator assemblies.

Acknowledgements Support for this research is greatly appreciated with funding through the Air Force Research Laboratory, Aerospace Component and Subsystem Division Titled: "Design, Model, and Fabrication of a 5x5 Large Tip, Tilt, and Piston MEMS Micromirror Array".

References

1. Yoo, B., Megens, M., Chan, T., Sun, T., Yang, W., Chang-Hasnain, C., Horsley, D., Wu, M.: Optical phased array using high contrast gratings for two dimensional beamforming and beamsteering. Opt. Express. **21**(10), 12238–12248 (2013)
2. Samuelson, S., Xie, H.: A large piston displacement MEMS mirror with electrothermal ladder actuator arrays for ultra-low tilt applications. J. Microelectromech. Syst. **23**(1), 39–49 (2014)
3. Sandner, T., Grasshoff, T., Schenk, H., Kenda, A.: Out-of-plane translatory MEMS actuator with extraordinary large stroke for optical path length modulation in miniaturized FTIR spectrometers. In: Sensor+Test Conferences 2011 – IRS^2 Proceedings, pp. 151–156 (2011)
4. Todd, S., Jain, A., Qu, H., Xie, H.: A multi-degree-of-freedom micromirror utilizing inverted-series-connected bimorph actuators. J. Opt. A Pure Appl. Opt. **8**, 352–359 (2006)
5. Wu, L., Dooley, S., Watson, E., McManamon, P.: A tip-tilt-piston micromirror array for optical phased array applications. J. Microelectromech. Syst. **19**(6), 1450–1461 (2010)
6. Jia, K., Pal, S., Xie, H.: An electrothermal tip-tilt-piston micromirror based on folded dual S-shaped bimorphs. J. Microelectromech. Syst. **18**(5), 1004–1015 (2009)
7. Lim, S., Choi, J., Horowitz, R., Majumdar, A.: Design and fabrication of a novel bimorph micro-opto-mechanical sensor. J. Microelectromech. Syst. **14**(4), 683–690 (2005)
8. Carter, J., Cowen, A., Hardy, B., Mahadevan, R., Stonefield, M., Wilcenski, S.: PolyMUMPs™ Design Handbook. (Revision 13). [On-Line]. http://memscap.com/__data/assets/pdf_file/0019/1729/PolyMUMPs-DR-13-0.pdf. Accessed 26 Jan 2018

D. Torres received a B.S. from the University of Puerto Rico at Mayagüez, and a Ph.D. from Michigan State University (MSU), both in electrical engineering. He is an electronics engineer in the Sensors Directorate at AFRL.

Chapter 15
Acoustic Emission Analysis on Mode I Delamination Tests of Fused Deposition Modelling Parts

C. Barile, C. Casavola, and A. Cazzato

Abstract The Fused Deposition Modelling (FDM), has become one of the most employed technologies to build complex 3D prototypes directly from a computerized solid model. In this process, the model is built as a layer-by-layer deposition of a feedstock wire. Moreover, the continuous improvements of the FDM technique have allowed to change the main purpose of this technique from rapid prototyping to a rapid manufacturing method. Anyway, the production of end-use parts strongly depends on the full understanding of the relationship between the process parameters and the mechanical behaviour. Consequently, it is fundamental to determine the FDM parts material properties as a function of the service load. The aim of the present work is to study the effect of the extrusion temperature on the inter-layer cohesion properties of FDM printed parts employing a double cantilever beam (DCB) test. According to ASTM D5528, the delamination energy G_I will be determined. Moreover, the delamination process will be investigated using acoustic emission technique and, in order to evaluate the crack opening, the whole test will be recorded by using a CCD camera calibrated with a digital grid.

Keywords Acoustic emission · Double cantilever beam · Fused deposition modelling · Mode I fracture mechanics · Orthotropic materials

15.1 Introduction

The Fused Deposition Modelling (FDM) is one of the most employed 3D printing techniques in both consumer and enterprise business. This process builds complex 3D prototypes starting from a computerized solid model. Currently, it has many fields of employing such as aerospace, medical, construction, and cultural [1, 2]. Moreover, the diffusion of the low-cost desktop 3D printers such as RepRap, Maker-Bot, Cube, Ultimaker, etc., has made this technology widely accessible at home and office. In this process, the part is built as a layer-by-layer deposition of a feedstock material in the form of a filament; it is partially melted, extruded and deposited by a numerically controlled heated nozzle [1]. After the deposition, the material cools, solidifies and sticks with the surrounding material and the nozzle can move upward, along the z-axis, for the deposition of the next layer. Once the entire model has been deposited, due to both the layer-by-layer approach and the orientation of the material laid, the FDM part shows orthotropic behavior like a laminate orthotropic structure [3–5]. Initially, the FDM printers could build parts only in acrylonitrile-butadiene-styrene (ABS) and polylactic acid (PLA). However, nowadays, many other materials have been developed and employed, e.g., ceramics, bio-resorbable polymer (PCL), metal, metal/polymers mixture materials, and short fiber composites. The PLA shows stronger mechanical resistance and lower thermal expansion than ABS. A low value of thermal expansion coefficient improves the printability of the material by reducing the de-layering phenomenon, the warp effect during the printing phase and the residual stresses problems [6–10].

The FDM technology has the great advantage to build single pieces or low volume products with tailored geometries and materials. In recent years, the continuous improvements of the FDM technology allowed to change the main purpose of this technique from rapid prototyping to a rapid manufacturing method [11]. However, the production of end-use parts strongly depends on the full understanding of the relationship between the mechanical behavior and the process parameters. Consequently, many studies are focusing on the analysis of mechanical properties of printed parts as a function of several printing parameters and mainly on the impact of the raster orientation on them [3, 12–15]. However, only few studies [16, 17] have explored the bond strength between layers and its key role on the strength of FDM parts. Reddy et al. [16], proved that the use of 90° rectangular testing specimens (ASTM D3039) was a good method to investigate the intra-layer bond strength of 3D-printed parts. Moreover, the authors highlighted the advantage, on the surface roughness, of depositing the beads of a

C. Barile (✉) · C. Casavola · A. Cazzato
Dipartimento di Meccanica, Matematica e Management (DMMM), Politecnico di Bari, Bari, Italy
e-mail: claudia.barile@poliba.it

M. Grady et al. (eds.), *Mechanics of Biological Systems & Micro-and Nanomechanics, Volume 4*, Conference Proceedings of the Society for Experimental Mechanics Series, https://doi.org/10.1007/978-3-319-95062-4_15

layer shifted compared to the previous layer. However, the influence of the modern design on the mechanical strength was not evaluated. Spoerk et al. [17] studied the effect of the layer-design on the mechanical properties of PLA 3D-printed parts aiming at maximizing both the intra- and inter-layer cohesion. However, despite a detailed parametric study, only the PLA material was investigated by Spoerk et al. [17].

In the present work, the effects of the extrusion temperature on the inter-layer cohesion properties of ABS parts will be studied by comparing mechanical tests and Acoustic Emission (AE) technique. Employing a DCB specimen, according to ASTM D5528 [18], three extrusion temperatures (i.e., 220, 230, 240 °C) will be investigated and the delamination energy G_I will be determined.

The use of AE approach is completely new for this kind of material and it aims to provide more useful information than the solely mechanical ones. Its purpose is to continuously monitor the state of material and, in particular, to identify both the nucleation and the propagation phases as well as all the events characteristics of the test. The main advantages connected with this technique derive from the nature of the method itself: it is a non-destructive passive technique, it doesn't require to produce events, but it only detects the internal changes which suddenly happen into the material subjected to both mechanical or thermal loads. It can detect rapid movements in stressed systems occurring from the largest scale (seismic events) to the smallest one (motion of dislocations). The possibility of fitting experimental data coming from a single event as well as from a succession of events has led researchers to develop scaled mathematical expression, successfully tested on industrial applications [19–21]. Along the last years, the large quantity of acoustic data collected combined with the variety of parameters to be analyzed (hits, events, amplitude, energy, location, wave forms, etc.) have provided useful information about the material behavior moving toward a deeply investigation by using statistical methods [22–26]. This approach has been successfully used for metals [27–29] and for Glass Fiber Reinforced Polymer [30, 31] but few results can be found in literature about its application to Carbon Fiber Reinforced Polymers [32, 33] and to Fused Deposition Modeling parts.

15.2 Materials and Method

15.2.1 Mechanical Tests

Tests were carried out on nine DCB specimens, three for each tested temperature. They are numbered from 1 to 3 for each group. Type codes A, B and C are associated with the extrusion temperatures and they refer to 220, 230 and 240 °C, respectively. Coupons are flat prismatic 125 mm long and having a cross section of 25 mm × 3 mm. To make an initiation site for the delamination, a film of Kapton was inserted at the midplane of each specimen during the printing phase. The insert thickness is less than 13 μm and is 63 mm long, according to the standard. The schematic of the sample with the load fixture is reported in Fig. 15.1.

According to ASTM D5528-01, a unidirectional printing strategy has been employed, i.e., the deposition of the infill is parallel to the major axes of the specimens. The samples have been printed by using a Ultimaker 2+, having 0.4 mm nozzle. Some parameters were kept constant for every specimen and these are reported in Table 15.1. The bed temperature was set to 90 °C and three extrusion temperatures were employed. Finally, the specimens have been fabricated with the minimum dimension perpendicular to the build platform.

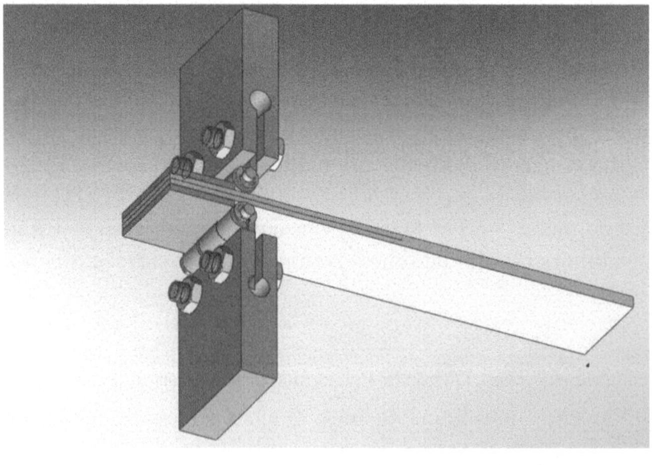

Fig. 15.1 Schematic of the sample with the load fixture used for mode I delamination test

Table 15.1 Fixed printer parameters

Parameter	Value
Air gap [mm]	0
Layer thickness [mm]	0.35
Bead width [mm]	0.70
Number of contour lines	3

Fig. 15.2 Image of one specimen tested by using digital grid for crack monitoring

In Table 15.1, the air gap represents the distance between two adjacent beads of the same layer; the layer thickness and the bead width are respectively the height and the width of a deposited filament. The number of contours represents the number of deposited edges before filling the inner part by infill beads.

Tests were carried out at constant displacement rate of 1 mm/min on an Instron 3343 electro-mechanical testing machine. Instead of creating a manual reference system to live monitor the crack growth as suggested in the ASTM D5528 standard [14], a digital grid was inserted and calibrated before each test. It has been overlapped on live recorded images acquired by CCD camera Marlin Allied (Fig. 15.2). The camera was calibrated and set so that images at 12.75 frame/s can be observed during the test. In order to facilitate monitoring grid is divided into two parts. The blue part corresponds to the length of the insert while the red part is used for evaluating the delamination propagation. According to [18] the blue grid, named a_0, represents the initial delamination length.

The delamination mechanism in a body characterized by linear elastic behavior provides a strain energy release rate (G) that is expressed by equation reported following:

$$G = -\frac{dU}{B\,da} \tag{15.1}$$

where a is the crack length, B is the specimen width and U represents the total potential energy of the specimen.

Energy release rate for Mode I DCB test is evaluated according to standardized formula referring to the Beam Theory (BT) as follows:

$$G_I^{BT} = \frac{3\,P\,\delta}{2\,B\,a} \tag{15.2}$$

where P is the maximum load and δ is the deflection in correspondence of the load. This theory may be corrected by considering the changing compliance at the end of the crack that causes both rotation and deformation of the specimen, modifying the distance between applied load and the end of the crack. This correction, known as Modified Beam Theory (MBT) reduces the overvalued G_I^{BT} toughness as follows:

$$G_I^{MBT} = \frac{3\,P\,\delta}{2B\,(a + \Delta)} \tag{15.3}$$

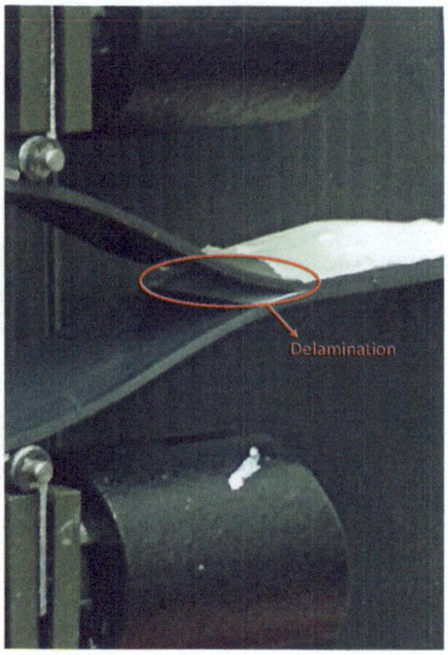

Fig. 15.3 Crack opening during delamination test

A representative image of delaminated specimen is reported in Fig. 15.3. It clearly shows the crack opening and propagation into the specimen at the end of insert.

15.2.2 Acoustic Emission

One piezoelectric sensor was placed on the sample's surface for detecting the acoustic signals. It was coupled to it by a silicone gel to avoid any type of signal reduction due to the presence of the air gap. The air acoustic impedance is, in fact, much lower than the ones of both material and sensor. The sensor used is a Pico sensor (5 mm × 4 mm), placed 10 mm far from the free edge of the specimen.

The major features of the Pico sensor are: 54 dB of peak sensitivity, operating frequency range between 200 and 750 kHz, resonant frequency of 250 kHz, temperature range between −65 and 177 °C. The signal was pre-amplified by a 20/40/60-AST preamplifier. The gain selector of the preamplifier was set to 40 dB.

The PCI-2 acquisition board was used to record the signals that were successively processed by the AEWin software. This package allows analyzing the key features of the signal such as hits, amplitude and energy. The acquisition frequency was set to 100 kHz while, after some preliminary tests, the threshold was set to 35 dB; at that level the noise is efficiently reduced without affecting the detection of the signal.

15.3 Results and Discussion

The analysis of results underlines a deep difference between specimens printed at the three temperatures (220, 230, 240 °C) according to three distinct aspects: the calculation of the G_I toughness by comparing the BT and the MBT, the different cracks opening, and the acoustic emissions signals detected. It is important to underline that for each group only two specimens were reported in the analysis process, due to the necessity of calibrating the optical parameters during the test recording.

Looking at the graph reported in Fig. 15.4, it is possible to observe the reduction of G_I values evaluated according to MBT respect to the BT one. This attitude confirms the large deformation characteristic of FDM specimens, especially if produced at higher extruded temperatures. The G_I value absolutely needs to be corrected by MBT respect to the one obtained by testing CFRP samples [33]. The deepest reduction of G_I toughness, in fact, refers to the C group produced at the highest temperature (240 °C).

The trend of the crack length related to the crosshead displacement is graphically reported for each specimen (Fig. 15.5). The different extrusion temperatures affect the crack behavior: for A group (220 °C) the lowest value of crosshead displacements is recorded corresponding to the highest value of crack length; for B group (230 °C) medium values of

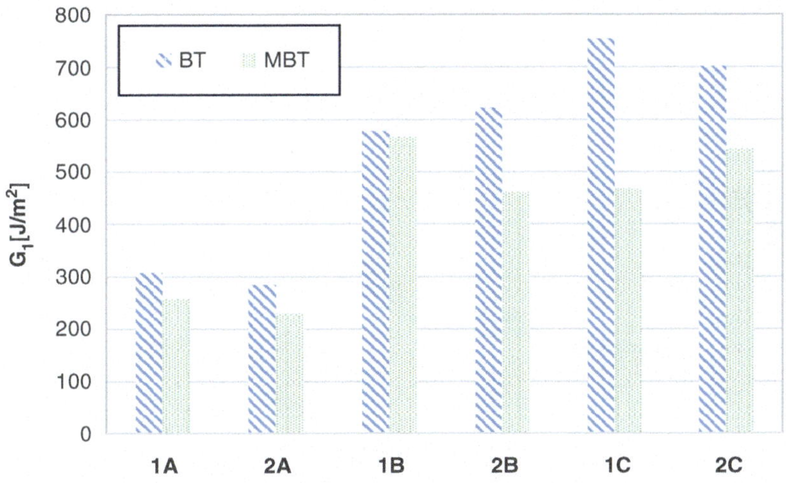

Fig. 15.4 Comparison of G_I values calculated with BT and MBT for all the three groups (A, B, C)

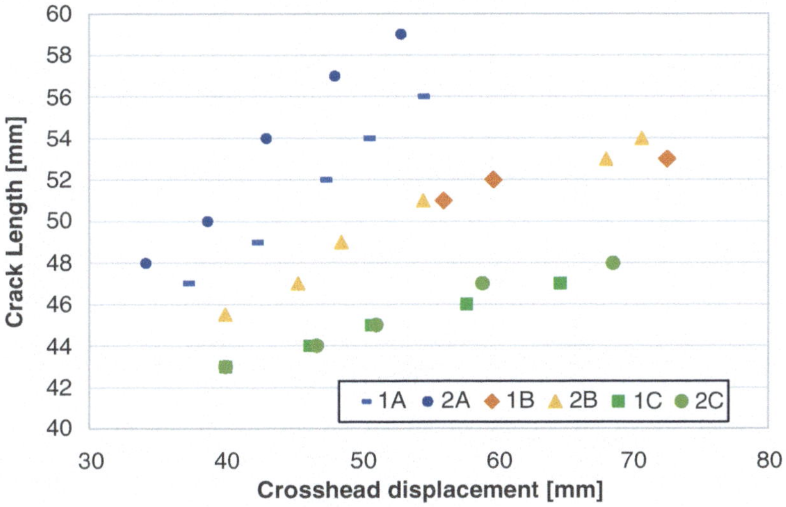

Fig. 15.5 Crack length VS Crosshead displacement of the three groups (A, B, C)

crack lengths are recorded and they are dispersed on a wide range of crosshead displacement; for C group (240 °C) the lowest values of crack length are reached even in correspondence of a large range of crosshead displacement. This attitude is justified by the different fusion level of layers connected to the different temperatures. It is correct to suppose that increasing the temperature increases the layers fusion and, as consequence, it results in highest values of G_I toughness.

Analyzing the Acoustic Emission data, key differences among the three temperatures could be observed too. Looking at the Fig. 15.6, in fact, it is possible to note two significant modifications: the first one refers to the number of Hits recorded, the second one relates to Amplitude values reached during all the tests of each group. The number of Hits seems to increase by increasing the extruder temperature and in the same way the highest Amplitude values appear for the group C, extruded at the highest temperature (240 °C).

This attitude suggests an increasing acoustic activity connected with the temperature increasing. Looking at the previous results this behavior indicates a growing deformation to delamination's disadvantage, and thus it indicates an increasing of G_I toughness. It could be assessed that the temperature increasing, supporting the layers fusion tends to compact the samples and led them to produce more acoustic signals due to the applied load.

But the most promising consideration that could be inferred from these graphs refers to the predictive attitude of the AE technique respect to the visual analysis of the mode I delamination tests. The latter indicates the onset of delamination in correspondence of the dashed lines reported. Even if the starting time of the delamination process differs from one group to another the predictive feature doesn't change. The AE approach can detect the starting of a critical state of the material many seconds before the dashed lines, suggested by a deeply increment of the Hits as well as the Amplitude values. This

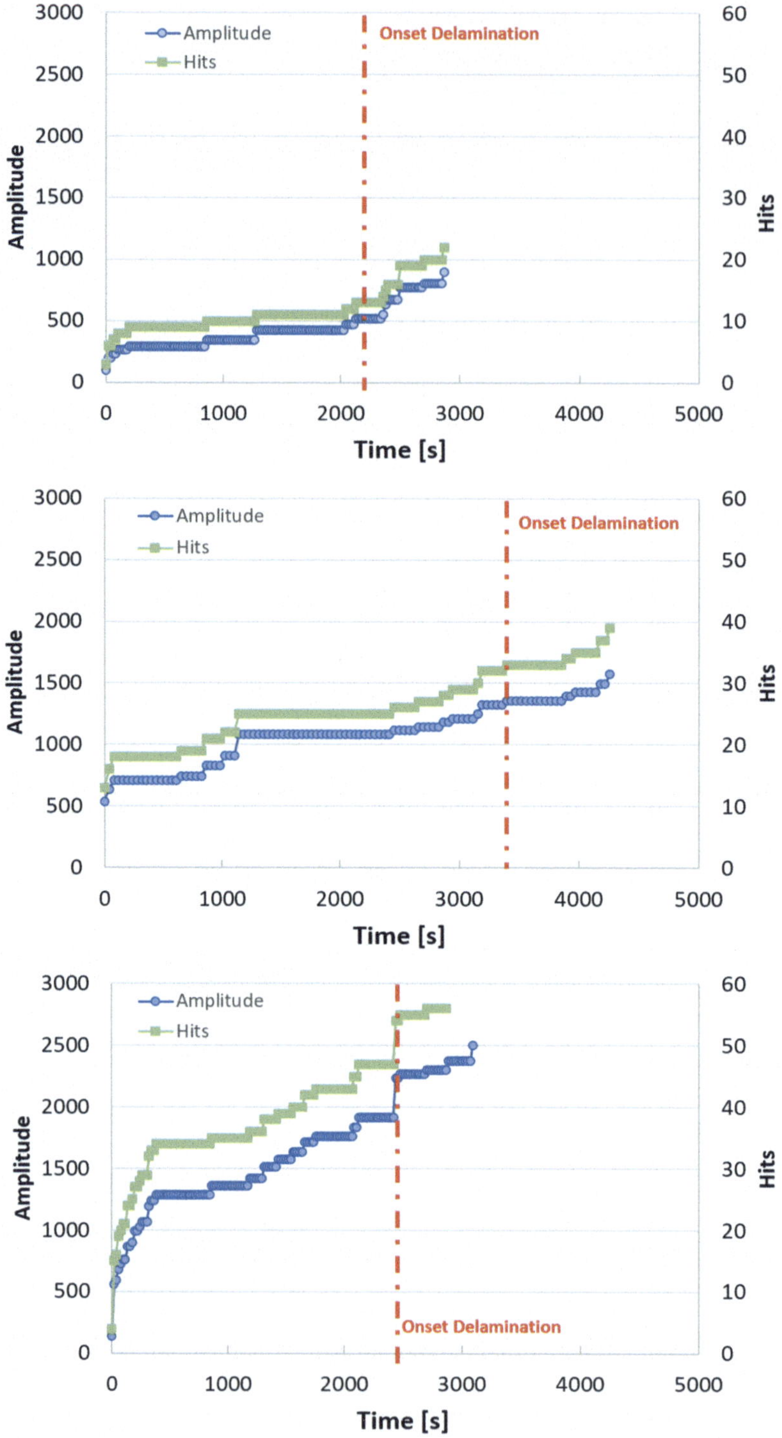

Fig. 15.6 Acoustic emission recorded data for the three groups (A, B, C)

attitude is more evident as the extruder temperature increases; in the third graph (group C) the deepest changes in slope occur, confirming a higher acoustic activity than in the other groups. This important advantage of AE, already assessed for other kind of material [27–33], appears extremely important for FDM parts, since they are not deeply affected by visible delamination but by internal structural modifications.

15.4 Conclusions

In this study the acoustic emission analysis is combined with the mode I delamination tests for mechanically characterized Fused Deposition Modelling parts, extruded at three different temperatures (220, 230, 240 °C). Results show that the temperature growing causes a reduced delamination respect to a large deformation phase, inducing to increase the G_I toughness. Acoustic emission data provide important predictive information about the state of the material, because it is able to indicate a stronger acoustic activity connected with both the highest extruded temperature and a range time previous to the visible delamination phase.

References

1. Yan, X., Gu, P.: A review of rapid prototyping technologies and systems. Comput. Aided Des. **28**(4), 307–318 (1996)
2. Petzold, R., Zeilhofer, H.F., Kalender, W.A.: Rapid prototyping technology in medicine—basics and applications. Comput. Med. Imaging Graph. **23**(5), 277–284 (1999)
3. Casavola, C., et al.: Orthotropic mechanical properties of fused deposition modelling parts described by classical laminate theory. Mater. Des. **90**, 453–458 (2016)
4. Barile, C., Casavola, C., Pappalettera, G., Pappalettere, C.: Hybrid characterization of laminated wood with ESPI and optimization methods. In: Conference Proceedings of the Society for Experimental Mechanics Series, vol. 3, pp. 75–83 (2013)
5. Barile, C., Casavola, C., Pappalettera, G., Pappalettere, C.: Mechanical characterization of SLM specimens with speckle interferometry and numerical optimization. In: Conference Proceedings of the Society for Experimental Mechanics Series, vol. 6, pp. 837–843 (2011)
6. Casavola, C., et al.: Preliminary study on residual stress in FDM parts. In: Residual Stress, Thermomechanics & Infrared Imaging, Hybrid Techniques and Inverse Problems, vol. 9, pp. 91–96. Springer, Cham (2017)
7. Casavola, C., et al.: Residual stress measurement in fused deposition modelling parts. Polym. Test. **58**, 249–255 (2017)
8. Casavola, C., Cazzato, A., Moramarco, V., Pappalettera, G.: Influence of printing constraints on residual stresses of FDM parts. In: Conference Proceedings of the Society for Experimental Mechanics Series, vol. 8, pp. 121–127 (2018)
9. Barile, C., Casavola, C., Pappalettera, G., Pappalettere, C., Tursi, F.: Drilling speed effects on accuracy of HD residual stress measurements. In: Conference Proceedings of the Society for Experimental Mechanics Series, vol. 8, pp. 119–125 (2014)
10. Barile, C., Casavola, C., Pappalettera, G., Pappalettere, C.: Residual stress measurement by electronic speckle pattern interferometry: a study of the influence of geometrical parameters. Struct. Integr. Life. **11**(3), 177–182 (2011)
11. Caulfield, B., McHugh, P.E., Lohfeld, S.: Dependence of mechanical properties of polyamide components on build parameters in the SLS process. J. Mater. Process. Technol. **182**(1–3), 477–488 (2007)
12. Ahn, S.-H., et al.: Anisotropic material properties of fused deposition modeling ABS. Rapid Prototyp. J. **8**(4), 248–257 (2002)
13. Croccolo, D., De Agostinis, M., Olmi, G.: Experimental characterization and analytical modelling of the mechanical behaviour of fused deposition processed parts made of ABS-M30. Comput. Mater. Sci. **79**, 506–518 (2013)
14. Letcher, T., Waytashek, M.: Material Property Testing of 3D-Printed Specimen in PLA on an Entry-Level 3D Printer: V02AT02A014 (2014)
15. Lužanin, O., Movrin, D., Plančak, M.: Effect of layer thickness, deposition angle, and infill on maximum flexural force in FDM-built specimens. J. Technol. Plast. **39**(1), 49–58 (2014)
16. Reddy, B.V., Reddy, N.V., Ghosh, A.: Fused deposition modelling using direct extrusion. Virtual Phys. Prototyp. **2**(1), 51–60 (2007)
17. Spoerk, M., et al.: Parametric optimization of intra-and inter-layer strengths in parts produced by extrusion-based additive manufacturing of poly (lactic acid). J. Appl. Polym. Sci. **134**(41), 45401 (2017)
18. ASTM D5528-01, Standard test method for mode I interlaminar fracture toughness of unidirectional fiber-reinforced polymer matrix composites. ASTM International, West Conshohocken (2007)
19. Grosse, C.U., Ohtsu, M.: Acoustic Emission Testing. Springer, New York (2008)
20. Anifrani, J.C., Le Floch, C., Sornette, D., Souillard, B.: Universal log-periodic correction group scaling for rupture stress prediction from acoustic emission. J. Phys. I France. **5**, 631–638 (1995)
21. Johansen, A., Sornette, D.: Critical ruptures. Eur. Phys. J. B. **18**, 163–181 (2000)
22. Rosti, J., Koivisto, J., Alava, M.J.: Statistics of acoustic emission in paper fracture: precursors and criticality. J. Stat. Mech. Theory Exp. **P02016**, 1–18 (2010)
23. Raischel, F., Kun, F., Herrmann, H.J.: Continuous damage fiber bundle model for strongly disordered materials. Phys. Rev. E. **77**, 046102 (2008)
24. Niccolini, G., Durin, G., Carpinteri, A., Lacidogna, G., Manuello, A.: Crackling noise and universality in fracture systems. J. Stat. Mech. Theory Exp. **2009**, P01023 (2009)
25. Varotsos, P.A., Sarlis, N.V., Skordas, E.S., Tanaka, H.K., Lazaridou, M.S.: Attempt to distinguish long-range temporal correlations from the statistics of the increments by natural time analysis. Phys. Rev. E. **74**, 021123 (2006)
26. Vallianatos, F., Michas, G., Benson, P., Sammonds, P.: Natural time analysis of critical phenomena: the case of acoustic emissions in triaxially deformed Etna basalt. Phys. A. **392**, 5172–5178 (2013)
27. Barile, C., Casavola, C., Pappalettera, G., Pappalettere, C.: Acoustic emission analysis of aluminum specimen subjected to laser annealing. In: Conference Proceedings of the Society for Experimental Mechanics Series 2014, vol. 8, pp. 309–315 (2014)
28. Barile, C., Casavola, C., Pappalettera, G., Pappalettere, C.: Hybrid thermography and acoustic emission testing of fatigue crack propagation in aluminum samples. In: Conference Proceedings of the Society for Experimental Mechanics Series 2015, vol. 66(5), pp. 247–252 (2015)

29. Barile, C., Casavola, C., Pappalettera, G., Pappalettere, C.: Analysis of crack propagation in stainless steel by comparing acoustic emissions and infrared thermography data. Eng. Fail. Anal. **69**, 35–42 (2015)
30. Saeedifar, M., Fotouhi, M., Najafabadi, M.A., Toudeshky, H.H., Minak, G.: Prediction of quasi-static delamination onset and growth in laminated composites by acoustic emission. Compos. Part B. **85**, 113–122 (2016)
31. Fotouhi, M., Najafabadi, M.A.: Acoustic emission-based study to characterize the initiation of delamination in composite materials. J. Thermoplast. Compos. Mater. **29**(4), 519–537 (2016)
32. Mohammadi, R., Saeedifar, M., Toudeshky, H.H., Najafabadi, M.A., Fotouhi, M.: Prediction of delamination growth in carbon/epoxy composites using a novel acoustic emission-based approach. J. Reinf. Plast. Compos. **34**(11), 868–878 (2015)
33. Barile, C., Casavola, C.: Fracture behavior of unidirectional composites analyzed by acoustic emissions technique. In: Conference Proceedings of the Society for Experimental Mechanics Series, vol. 7, pp. 121–127 (2018)

Chapter 16
Experimental Study of the Mechanics of Blast-Induced Traumatic Brain Injury

Joseph Kerwin, Suhas Vidhate, Faezeh Masoomi, Michaelann Tartis, Adam M. Willis, and Ricardo Mejia-Alvarez

Keywords Blast induced traumatic brain injury · Improvised explosive device · Gelatin brain phantom · Post traumatic stress disorder · Particle image velocimetry

16.1 Introduction

Blast-induced traumatic brain injury (bTBI) has become a "signature wound" of modern combat due to the expansive use of improvised explosive devices (IED). bTBI is characterized as either primary, secondary, tertiary, or quanternary. Primary bTBI is characterized when a shock wave causes damage directly to the biological material (e.g., brain tissue) while propagating through the skull. Whereas, secondary and tertiary bTBI involves damage caused by matter thrown by the explosion or the result of a person being physically moved by a shock wave, respectively. Over 60% of combat casualties haven been caused by explosive blasts from the operations in Iraq and Afghanistan [2, 3]. However, while our understanding of trauma and resuscitative techniques have improved, thereby reducing combat mortality, the same improvements have not been found in brain injury; thus, the incidence of neurologic morbidity has increased.

This lack of improvement stems from the fact that the mechanisms that cause bTBI still remain poorly understood. Although secondary and tertiary injury may have analogous mechanisms with traditional traumatic brain injury, the mechanisms of primary blast injury are thought to be both unique and mechanically very complex to characterize. The complexity stems from the challenges of shock wave physics, complex anatomy of the human cranial cavity, the complex material properties of biologic tissue under high strain rates, and the sub-millisecond timescale of injury.

With the purpose of understanding the mechanics of bTBI, we used controlled blasts to load bio-fidelic cranial phantoms based on organic gelatins. Unlike animal models, phantoms allow for better visualization of intracranial contents, appropriate length scales without the need for complex and invalidated mechanical scaling, and increased reproducibility while reducing the cost, time, and ethical issues. Four gelatin-based materials were used in the creation of the cranial phantoms for our experiments. White and gray matter have been mimicked by tailoring the mechanical properties to match that of neural tissue with bovine gelatin along with different weight concentrations of Polyacrylamide (PAA).

16.2 Methods

Due to the possibility of matching a wide range of material properties with PAA, phantoms fabricated with this material are advantageous compared to phantoms fabricated with other gelatins. Given that PAA is translucent in nature, it allows for the implementation of optical imaging capabilities. Also, PAA can be poured and cross-linked at room temperature, saving time and energy costs during fabrication. The fabrication process followed herein builds on previously reported methods [1, 4, 5].

J. Kerwin (✉) · S. Vidhate · F. Masoomi · R. Mejia-Alvarez
Department of Mechanical Engineering, Michigan State University, East Lansing, MI, USA
e-mail: kerwinj1@msu.edu

M. Tartis
Department of Chemical Engineering, New Mexico Institute of Mining and Technology, Socorro, NM, USA

A. M. Willis
Department of Mechanical Engineering, Michigan State University, East Lansing, MI, USA

Department of Neurology, San Antonio Military Medical Center, Fort Sam Houston, TX, USA

M. Grady et al. (eds.), *Mechanics of Biological Systems & Micro-and Nanomechanics, Volume 4*, Conference Proceedings of the Society for Experimental Mechanics Series, https://doi.org/10.1007/978-3-319-95062-4_16

Fig. 16.1 Simulated test object for blast testing

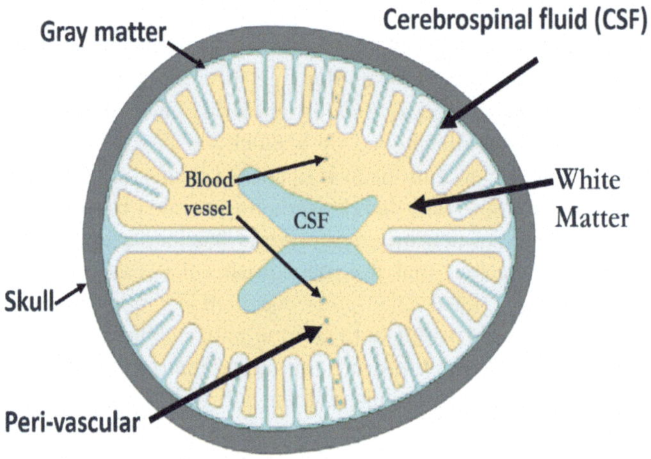

Fig. 16.2 2D representation of simulated test objects with descriptions

The gelatin simulants have been molded to a simplified geometric representation based on an MRI axial section of a real human head. Relative size and shape of a human head was maintained. The phantoms incorporated only the dominant length scales of the gyri and sulci of the original MRI image, as well as cylindrical cavities to simulate blood vessels. These features can be seen in Fig. 16.1. To our knowledge, never before have such complex gelatin brain phantoms ever been used in experimental blast loading.

To model the intracranial environment, each gelatin phantom was made slightly smaller than its skull, and the resulting internal gap ($\sim$ 2 mm) filled with a cerebrospinal fluid (CSF) simulant. This CSF simulant was made by adding albumin to deionized water. As can be seen in Figs. 16.1 and 16.2, the global geometry of the test object is an extrusion of an axial cut of the brain, or roughly an oblong cylinder. Its two opposing flat bases serve as optical windows for illumination and visualization. Each brain phantom was seeded with either glass micro-spheres, nano-bubbles, silica carbide particles, or a combination of the three, in a strategically located transverse plane (either half the way or a quarter of the way between the flat ends). The particles embedded in this transverse plane were used to track local displacement as the phantom is loaded with blast waves and the induced motion evolves afterward.

Full scale shock wave interaction of the biofidelic gelatin phantoms took place $\sim$1 m inside of a blast tube. The blast tube design was 21.3 m long with a 0.61 m radius. Pentaerythritol tetranitrate (PETN) was used to create the blast wave for all experiments. Over pressures ranged between 235 and 450 kPA. Phantoms were blast-loaded under two configurations: Frontal and Temporal. The former means the "frontal lobe" of the phantom facing the incoming wave, and the latter means one of the "temporal lobes" of the phantom facing the incoming wave.

High speed imaging was used to visualize blast wave propagation and its immediate aftermath within the phantom. To this end, four ultrahigh-speed digital cameras (Phantom V2512 Series) where focused on the same field of view by means of a beam-splitter array. With this configuration, the cameras were triggered in sequence to effectively obtain a frame rate four times as fast as a single camera. This strategy allowed us to achieve a time resolution of 10 μs at full spatial resolution

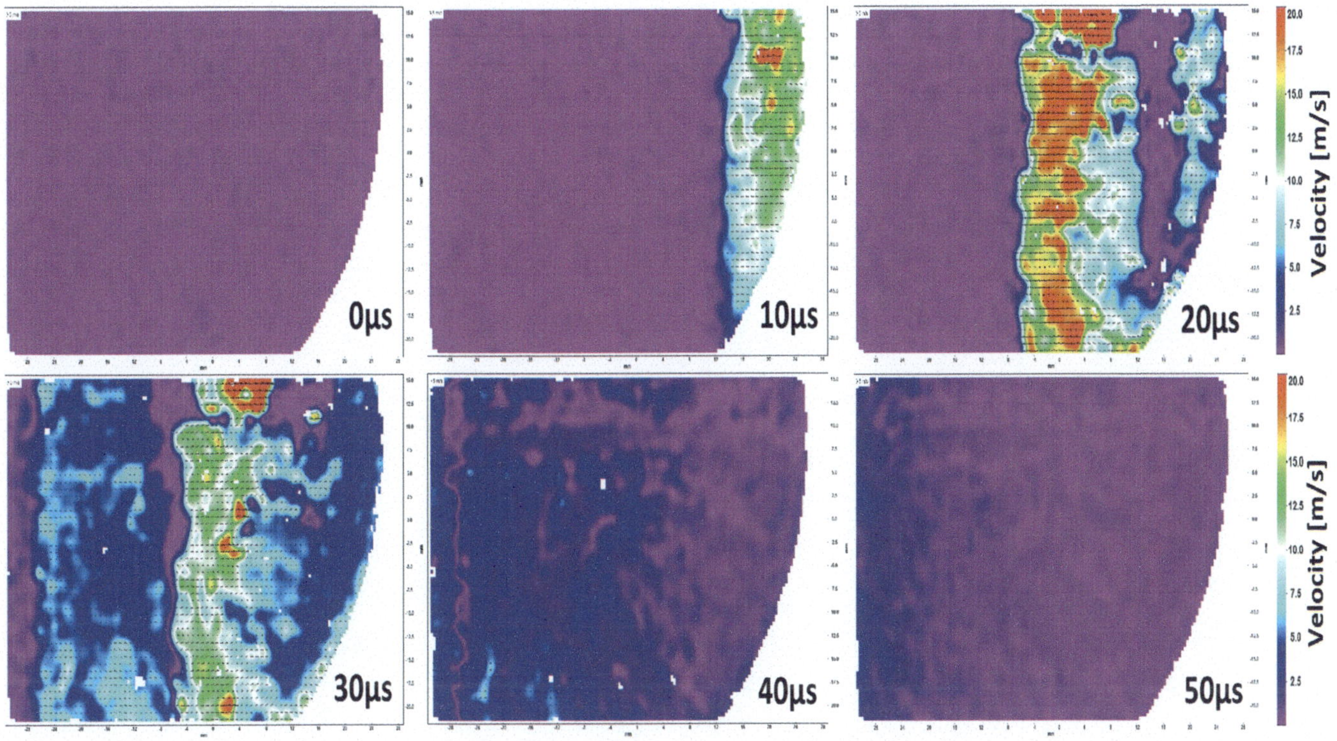

Fig. 16.3 Forward propagation of blast wave through coup configuration of cranial phantom

(1280 pixel × 800 pixel). An array of high-intensity LED's were used to continuously illuminate the test object during blast loading. DaVis (LaVision 8.3 software) was used to record the images in real time and post-process the data using particle image velocimetry (PIV). For each experimental run, the field of view was around 4 cm × 4 cm, which was in fact a zoomed-in view of a portion of the test object to obtain high spatial resolution in the areas of interest. These areas are the *coup* (portion of the brain facing the incoming wave) and the *contre-coup* (portion of the brain directly opposite to the incoming wave). These are known to be the regions that receive most of the loading, and hence, where most of the localized damage happens.

16.3 Discussion

As demonstrated in Fig. 16.3, particle displacement vectors are showing a shock wave's initial pass through the gelatin phantom. As compared to traditional PIV, where particle mapping is estimated through consecutive images of particle displacement, we calculated absolute particle displacement with respect to a common reference. This was achieved by first creating a time series of the blast propagation through the gelatin simulant based on the original reference position. Our method allows for not only particle position but also the relative magnitude for each particle displacement. By using the instantaneous local displacement the strain, strain-rate, and all associated kinematics of deforming solids can be determined using PIV algorithms.

References

1. Cao, R., Huang, Z., Varghese, T., Nabi, G.: Tissue mimicking materials for the detection of prostate cancer using shear wave elastography: a validation study. Med. Phys. **40**(2), 022903 (2013)
2. Elder, G., Mitsis, E., Ahlers, S., Cristian, A. Blast-induced mild traumatic brain injury. Psychiatr. Clin. N. Am. **33**(4), 757–781 (2010)
3. Ling, G., Bandak, F., Armonda, R., Grant, G., Ecklund, J.: Explosive blast neurotrauma. J. Neurotrauma **26**(6), 815–825 (2009)

4. Sun, M., Shieh, J., Lo, C., Chen, C., Chen, B., Huang, C., Chen, W.: Reusable tissue-mimicking hydrogel phantoms for focused ultrasound ablation. Ultrason. Sonochem. **23**, 399–405 (2015)
5. Zell, K., Sperl, J., Vogel, M., Niessner, R., Haisch, C.: Acoustical properties of selected tissue phantom materials for ultrasound imaging. Phys. Med. Biol. **52**(20), N475–N484 (2007)

Chapter 17
Study on the Fabrication Process of a MEMS Bistable Energy Harvester Based on Coupled Component Structures

Masoud Derakhshani, Brian E. Allgeier, and Thomas A. Berfield

Abstract Due to the rapid growth in demand for power for sensing devices located in remote locations, scientists' attention has been drawn to vibration energy harvesting as an alternative to batteries. As a result of over two decades of micro-scale vibration energy harvester research, the use of mechanical nonlinearity in the dynamic behavior of the piezoelectric power generating structures had been recognized as one of the promising solutions to the challenges presented by chaotic, low-frequency vibration sources found in common application environments. In this study, the design and performance of a unique MEMS-scale nonlinear vibration energy harvester based on coupled component structures and bi-stable states are investigated. The coupled-components within the device consist of a main buckled beam bonded with piezoelectric layers, a torsional rod, and two cantilever arms with tip masses at their ends. These arms are connected to the main beam through the torsional rod and are designed to help the main beam snap between its buckled stable states when subjected to sufficient vibration loading. The fabrication of the device will be discussed, including use of plasma-enhanced chemical vapor deposition (PECVD) of silicon nitride under an alternating power field to control compressive stress development within the main buckled beam. After completing the fabrication process, the next step would be testing the device under a variety of vibration loading conditions for its potential use as a vibration energy harvester.

Keywords Vibration energy harvesting · MEMS · Bistability · Compressive buckled beam · Experimental fabrication

17.1 Introduction

Vibration energy harvesting is a well-known method to harvest electrical energy from ambient mechanical motions. For linear systems, this method is firmly rooted in the typical equations of motion for mechanical vibration, which leads such systems to work efficiently only at a narrow bandwidth close to their natural frequencies. It was found that using physical properties of mechanical nonlinearity could be a good solution to this issue and significantly improve both the frequency bandwidth and the output power of energy harvesters. There have been a lot of works done in this field to show the nonlinear behavior of post-buckling systems in micro-scale structures [1, 2].

There are numerous studies on MEMS vibration energy harvesters to improve their performance and make them work in real-life applications. Fang et al. [3] designed and fabricated a MEMS-scale piezoelectric-based vibration energy harvester using different techniques of micro fabrication. Their experimental testing of the fabricated devices showed that improved microfabrication methods could produce good performance compared to typical micro-scale vibration energy harvesters. Liu et al. [4] developed a MEMS-scale piezoelectric based power generator consisting of cantilevers array to improve the frequency range and power output. The structural material for the cantilever was silicon with PZT film as a transducer and a proof mass is attached to the free end of each beam. The experimental results showed that the arrayed design of cantilevers with different dimensions makes the system work at a broader range of frequencies and ultimately is able to generate more output power. Ferrari et al. [5] studied the efficiency enhancement of a nonlinear piezoelectric vibration energy harvester coupled with a permanent magnet. A piezoelectric bimorph cantilever was considered as the harvester. The tip mass of the cantilever was faced towards a fixed permanent magnet, which serves a role in creating bi-stability in the system. The experimental results showed that for a low enough distance between the tip mass and the magnet, a softening behavior could be captured which shifts the resonances towards lower frequencies. It was also observed that snap-through motion can be created when the system is excited by white-noise mechanical vibrations. In 2015, Rezaeisaray et al. [6] designed and fabricated a multi-degree of freedom system as a vibration energy harvester functional at low excitation frequencies. In their

M. Derakhshani (✉) · B. E. Allgeier · T. A. Berfield
University of Louisville, Louisville, KY, USA
e-mail: masoud.derakhshani@louisville.edu

M. Grady et al. (eds.), *Mechanics of Biological Systems & Micro-and Nanomechanics, Volume 4*, Conference Proceedings of the Society for Experimental Mechanics Series, https://doi.org/10.1007/978-3-319-95062-4_17

proposed design, a big proof mass is connected to the device frame by two cantilever beams. The structure is fabricated from a silicon substrate, and aluminum nitride was used as the electromechanical converter material. One of the advantages pointed out in this work was being able to reduce the first three natural frequencies of the micro-scale size of the system to within the range of ambient vibrations. The potential of a flexible structure consisting of multiple beams connected with joint masses in a zigzag combination as a vibration energy harvester was studied by Zhou et al. [7]. In this study, a theoretical model was proposed and compared with both finite element analysis and experimental results. The obtained results showed that the resonant frequencies of the system can be significantly decreased by increasing the number of the considered beams and, consequently, can be more practical for low-frequency ambient vibrations. Another advantage was pointed out in this research is the functionality of this device under multi-directional excitations, which makes it a suitable approach to harvest energy from human body motion.

While a lot of studies have been done so far on the improvement of functionality of micro-scale vibration energy harvesters, the absence of a solid promising design to address the above-mentioned issues is still needed in this area. In this study, steps have been taken towards fabricating a unique-designed MEMS structure as a vibration energy harvester, which has a very high potential to resolve the literature mentioned problems and work efficiently at ambient vibration frequency levels.

17.2 Model Description and Experimental Process

A computer model of the proposed MEMS-scale device described in this research is shown in Fig. 17.1, consisting of a main buckled beam, in which the buckling created by a compressive load. This beam is connected to two cantilever arms via a clamped-ends rod which role is transferring motion between the main buckled beam and these two arms. In order to tune the device flexibility and lowering its natural frequencies, a tip mass is considered at the free end of each arm. Different colors of the top view in this figure is related to different layers of the device including structural layer, bottom electrode, piezoelectric layer and top electrode. Dynamic behavior of a macro-scale version of this structure has been analyzed in [8] and it was shown that the dynamic performance of the structure is significantly dependent on different parameters including the level of compressive load as well as the geometry of different components and for certain cases, it is possible to capture snap-through motion, which is crucial for its functionality as a vibration energy harvester.

The fabrication of the proposed device requires five photomasks corresponding to patterning of the following material layers or processes: stressed silicon nitride, bottom electrode, piezoelectric, top electrode, and the final bulk etching step. These masks are displayed in Fig. 17.2 superimposed on top of one another. The main structural layer of the device is made of stressed nitride, on which the electrode and piezoelectric layers are deposited. All microfabrication processes discussed in this research have taken place at the University of Louisville's Micro/Nano Technology Center (MNTC) [9].

All fabrication processes of this device were performed on a four-inch silicon wafer. The Si wafer maintains a thickness of 380 μm and (100) crystalline orientation. The main structural layer considered for this device is stressed nitride. This material is the first fabricated layer supplies the support for the top layers of the device. Furthermore, the controlled stress created in this layer provides the required buckling load for the main beam. This layer is fabricated using plasma-enhanced chemical vapor deposition machine (PECVD) at the University of Louisville's Micro/Nano Technology Center.

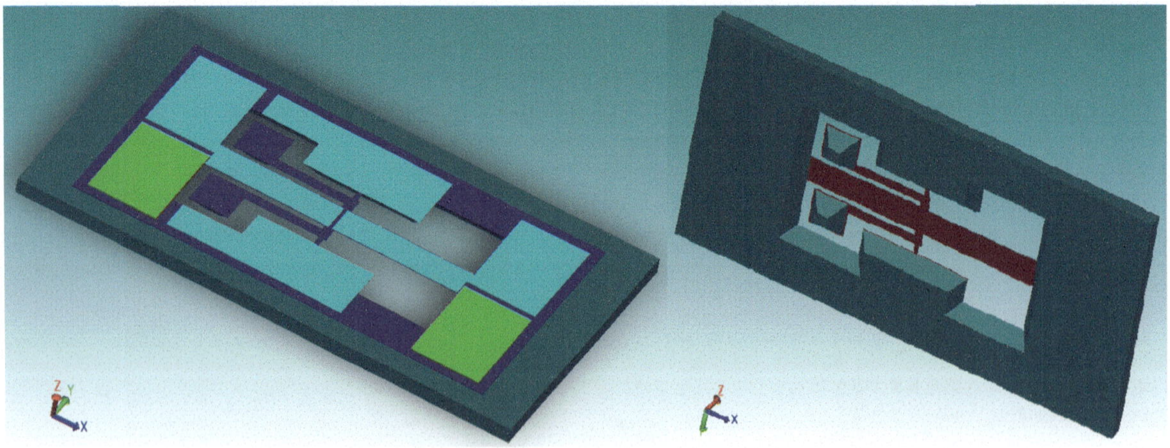

Fig. 17.1 Schematic design of the proposed MEMS vibration energy harvester; left figure: top view; right figure: bottom view

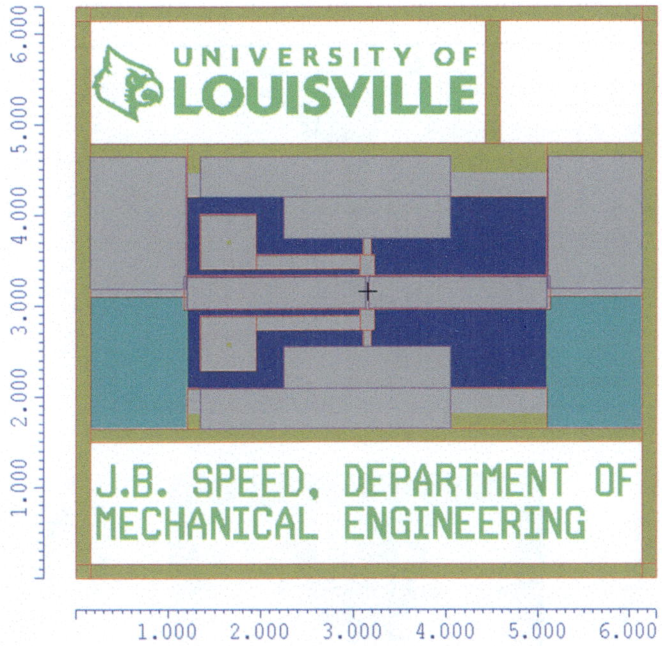

Fig. 17.2 Superimposed layouts of all five device photomasks: bulk etching (blue), stressed nitride (green), bottom electrode (cyan), piezoelectric (red), top electrode (violet)

Molybdenum is considered as the metal material for both the bottom and top electrodes. These two layers are electrically connected to the top and bottom face of the piezoelectric layer, which role is generating electrical current from the induced strain in the main structure created by mechanical vibrations. All metal layers were deposited via physical vapor deposition (PVD) process and were patterned with their own masks in a photolithography process, followed by a dry etching step with a Trion Metal Etcher tool. The piezoelectric material is deposited and patterned between two electrode layers, which is responsible for harvesting electrical current from the vibration load. The material considered for this part is aluminum nitride. The deposition process was taken the same way as mentioned in [10] by using PVD sputtering machine. In this method, the target chosen for the sputtering process is aluminum which was combined with a certain portion of nitrogen and argon gasses.

The last step of the fabrication process is backside etching, which releases the whole structure by etching away the bottom silicon layer of the wafer and turned out to be a little challenging due to the type of etching process had to be chosen for this specific device. Because of some certain required rules of the UofL Micro/Nano Technology Center, it has been decided to use anisotropic wet etching step for releasing the structure from its substrate. The etching technique was used in this research is potassium hydroxide (KOH) wet etching which etches straight walls at a 54.7° angle. A software named ACES (Anisotropic Crystalline Etching Simulator) was used to simulate the whole process of bulk etching and based on the software calculation, the required etchant solution for the considered wafer thickness and orientation is 45% of KOH.

17.3 Results and Discussion

All the surface micro-fabrication processes discussed in the previous section were successfully performed step by step and a picture of the wafer after fabrication of the top electrode layer is shown in Fig. 17.3. The piezoelectric effect of the deposited aluminum nitride layer was successfully tested by an oscilloscope and it was shown some noticeable spikes when the wafer sample was lightly struck, creating a small amount of strain.

The challenging part of the fabrication is the very last step, which releases the whole structure from its substrate and has been tried to be done by KOH wet etching technique. Unfortunately, the results obtained for stressed nitride layer were not the expected ones and the moment that the structural layer is released, the device could not be held together and bursts after the etching process. This phenomenon can be due to multiple reasons. One of them is the amount of induced stress in the structure which is high enough to break the whole top layers when they lose their support from the substrate. In order to check this possibility, a test was performed on a wafer with a patterned nitride layer via unstressed nitride recipe. The

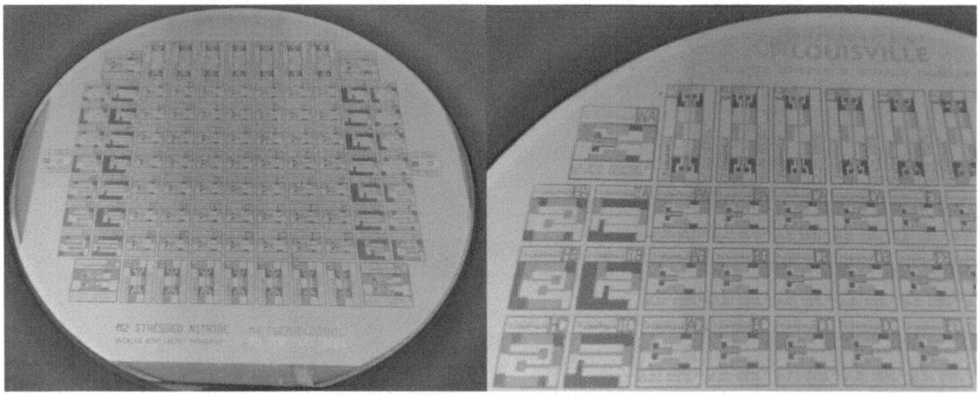

Fig. 17.3 A processed wafer after the deposition and patterning of the tope electrode layer

Fig. 17.4 Several examples of successfully released devices for unstressed nitride layer

mentioned release process was applied on the test wafer to free the unstressed nitride layer. It was shown that the nitride layer could be successfully freed in many cells without any major damage (Fig. 17.4).

Another reason may cause such thing to happen is the lack of rigidity of the structural layer due to not having enough thick stressed nitride layer on the top. To tackle this issue, it needs to increase the time deposition of stressed nitride via PECVD machine. However, increasing the thickness of the structural layer could lead to a less flexible buckled beam and consequently make it hard for the device to switch between its stable states as it is expected to do. One more thing needs to be considered in this study is the method chosen for releasing the structure from the backside substrate. Although the etching parameters were chosen based on our precise calculations from the simulation results, over-etching phenomenon, as well as the possibility of leaking KOH solution into the wafer fixture and destroying the unexposed side of the processed wafer, were two major problems created during the experimental process. One way to resolve this issue is using deep reactive-ion etching method (DRIE) instead of wet KOH etching, which can result in more controlled process and accurate geometries and consequently better output.

17.4 Summary and Conclusion

In this study, design and fabrication of a micro-scale bi-stable energy harvester consisting of different coupled components are thoroughly investigated and different steps of the fabrication process are discussed in details. As mentioned in the previous sections, it has been achieved very successful results for the surface fabrications of different layers of the device as expected. However, the very last step of bulk etching was failed to perform by KOH wet etching method, which made it impossible to dynamically test the device for its functionality. There are multiple reasons for this issue, including the large compressive stress created in the nitride layer and the lack of rigidity of the device due to too much thinness of the top deposited layers.

While these mentioned issues can be resolved in the fabrication process by further experiments, there needs to be a balance between the main beam flexibility and the induced stress of the structural layer, so the output device is able to satisfy the expected effects of the desired bi-stable vibration energy harvester without being damaged during the fabrication process.

References

1. Harne, R.L., Wang, K.W.: A review of the recent research on vibration energy harvesting via bistable systems. Smart Mater. Struct. **22**, 023001 (2013)
2. Ghasemi, A., et al.: Analytical analysis of buckling and post-buckling of fluid conveying multi-walled carbon nanotubes. Appl. Math. Model. **37**(7), 4972–4992 (2013)
3. Fang, H.-B., et al.: Fabrication and performance of MEMS-based piezoelectric power generator for vibration energy harvesting. Microelectron. J. **37**(11), 1280–1284 (2006)
4. Liu, J.-Q., et al.: A MEMS-based piezoelectric power generator array for vibration energy harvesting. Microelectron. J. **39**(5), 802–806 (2008)
5. Ferrari, M., et al.: A single-magnet nonlinear piezoelectric converter for enhanced energy harvesting from random vibrations. Sens. Actuators A Phys. **172**(1), 287–292 (2011)
6. Rezaeisaray, M., et al.: Low frequency piezoelectric energy harvesting at multi vibration mode shapes. Sens. Actuators A Phys. **228**, 104–111 (2015)
7. Zhou, S., et al.: Analytical and experimental investigation of flexible longitudinal zigzag structures for enhanced multi-directional energy harvesting. Smart Mater. Struct. **26**(3), 035008 (2017)
8. Derakhshani, M., Berfield, T., Murphy, K.D.: Dynamic analysis of a bi-stable buckled structure for vibration energy harvester. In: Dynamic Behavior of Materials, vol. 1, pp. 199–208. Springer, Cham (2018)
9. University of Louisville: Micro/Nano Technology Center.http://louisville.edu/micronano
10. Hwang, B.-H., et al.: Growth mechanism of reactively sputtered aluminum nitride thin films. Mater. Sci. Eng. A. **325**(1–2), 380–388 (2002)

Masoud Derakhshani is a Ph.D. candidate at the Mechanical Engineering Department of University of Louisville in Louisville, KY. His research is focused on the dynamic modeling and fabrication of a unique-designed MEMS-scale vibration energy harvester.

Chapter 18
Development of Biofilm-Surface Adhesion Technique via Laser-Induced Stress Waves

James D. Boyd, Stuart C. Ross, and Martha Grady

Abstract Bacterial infection is of paramount concern regarding human health and is a substantial hurdle in dental implantation. Bacterial proliferation on the surface of a dental implant can lead to development of a biofilm, a scenario in which bacteria have congregated and secrete an extracellular polymeric substance (EPS). The EPS has different modes of function, those most existential to the bacteria are: strong attachment to surfaces, a protective realm for the bacteria to flourish, containment of DNA, and a medium for quorum sensing (communication method used to recruit more bacteria). If this bacterial aggregation continues it can lead to serious subgingival infections which are difficult to remedy and often result in extraction of the implant to treat the site. Development of surfaces which are impervious to bacterial adhesion is of interest to the dental industry as such an accomplishment could greatly reduce the number of failed dental implants and therefore promote oral health and hygiene. To approach this problem, an evaluation of oral bacteria's adhesion strength to a given surface must be established. Laser spallation is an effective strategy well suited for this study, the technique involves loading a substrate with a stress wave delivered by an infrared laser. The substrate is designed with the same parameters as a dental implant and hosts a live biofilm comprised of an oral bacterium that is most common on failed implants. This work will describe the substrate preparation, biofilm selection, fluence to failure testing, and obstacles involved.

Keywords Biofilms · Laser spallation · Adhesion · *Streptococcus mutans* · Dental implants

A specific strain of bacteria must be selected based on specific characteristics, the most relevant being early colonization. Early colonizing bacteria is key to biofilm formation because these microbes are first to establish a domain on a fresh implant. The initial bacterial colonization process is dominated by oral Streptococcus, which make up over 80% of the early biofilm constituents [1]. The substrate that is inoculated with live bacteria must be the same material as an industry standard implant. Once the appropriate strain of bacteria has been selected and successfully cultured on a substrate that is representative of a dental implant, the sample assembly can then be used in an experiment that determines the sample-biofilm interfacial adhesion. If successfully executed, data collection and analysis of this type could enhance the biocompatibility of implant material and alleviate complex problems associated with installation of synthetic devices in the body [2].

Thin film adhesion studies have been conducted in the past through a multitude of techniques, some are more desirable than others depending on the film of interest and goal of research [3, 4]. Some of the classic film adhesion test methods are scratch, peel, stud-pull, four-point bending, topple tests, etc. A risk to consider when using these tests is unintentional damage to the sample as all these tests involve some form of direct contact. In these tests the interface is subject to very high stress levels and subsequently, non-uniform deformation occurs. As a result, data from these tests tend to be difficult to interpret and the results are qualitative and comparative [5].

Laser spallation acts as an adhesion strength test by delivering a photoacoustic shockwave to the biofilm with intent of ablating the film from its applied surface. Use of laser spallation to determine fluence of spallation is followed by a calibration procedure where the amount of stress created at the point of film detachment is measured. The laser spallation method is superior to other film adhesion testing methods as it is a non-contact method which targets only a specific portion of the sample leaving the rest of the area unscathed and available for further experimentation. The laser spallation setup used is illustrated in Fig. 18.1a. Laser spallation has been used in the past to quantitatively measure varying different thin film adhesion strengths [6–9] including mammalian cells [10–13] but not yet on biofilms.

J. D. Boyd (✉) · S. C. Ross · M. Grady
Department of Mechanical Engineering, College of Engineering, University of Kentucky, Lexington, KY, USA
e-mail: jdbo227@g.uky.edu

M. Grady et al. (eds.), *Mechanics of Biological Systems & Micro-and Nanomechanics, Volume 4*, Conference Proceedings of the Society for Experimental Mechanics Series, https://doi.org/10.1007/978-3-319-95062-4_18

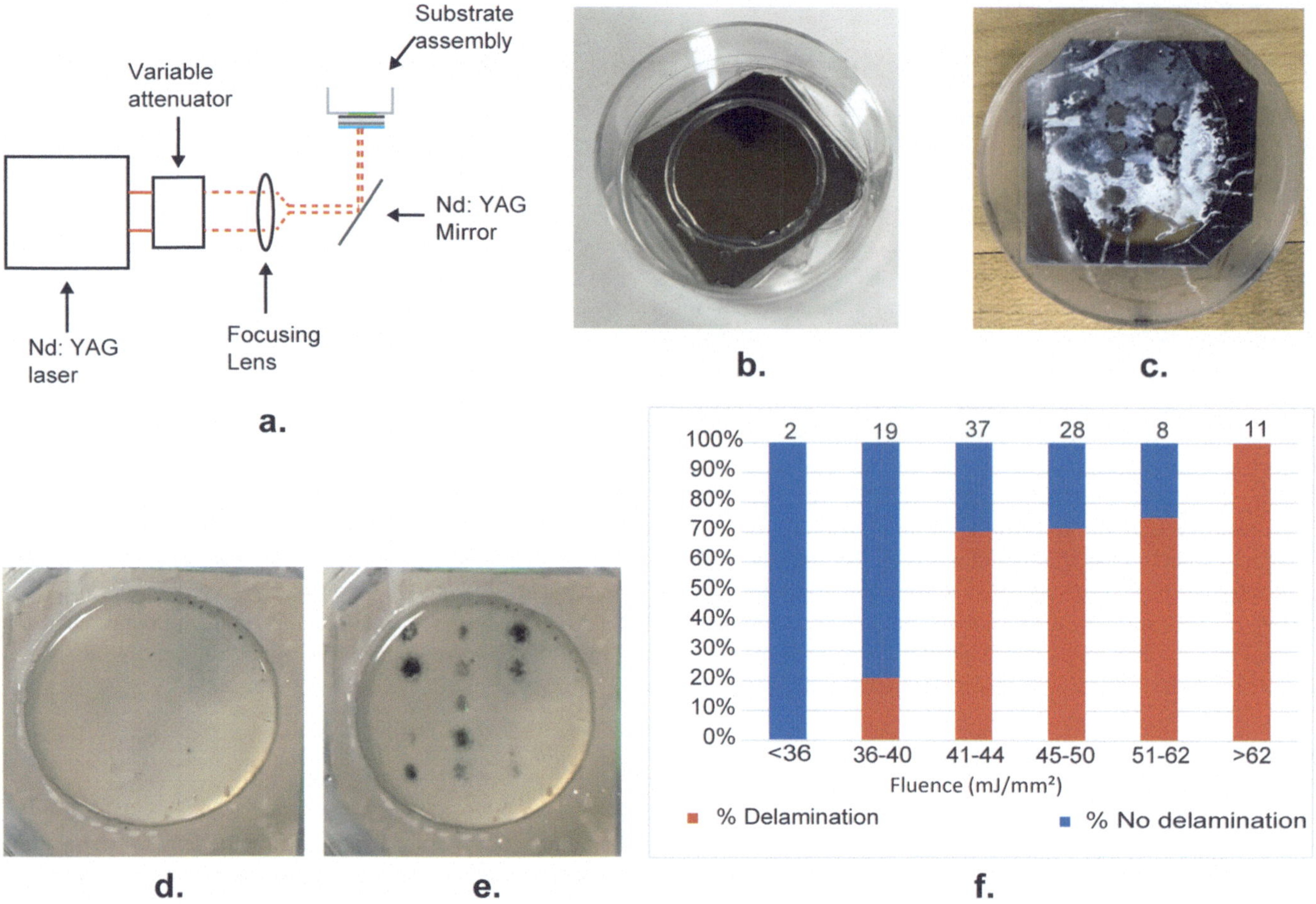

Fig. 18.1 (**a**) The laser spallation set up used during delamination experiments. The Nd: YAG laser used reflected 90° to allow for horizontal orientation of the substrate assembly used. (**b**) Shows the typical substrate assembly used, the titanium where the bacteria will be cultured is shown here, while the aluminum energy absorbing layer is on the other side not picture. (**c**) Water-glass layer that has been disrupted and is no longer uniformed due to the parafilm contacting the surface during incubation of bacteria. *S. mutans* biofilm grown on the titanium surface is shown in (**d**) while the same sample that has been loaded is shown in (**e**). The data collected for the experiments performed was combined into a single graph to show the general trend of how fluence impacts delamination of biofilm (**f**)

Streptococcus mutans was selected as an initial bacteria for this assay as it is known to be a main constituent of dental plaque, and is commonly linked to dental cavities [14]. The substrate selection for this experiment was driven by material and dimensional constraints. The material constraints controlled the accuracy of the experiment in its entirety because the substrate had to represent a dental implant. The industry standard dental implant is a commercially pure titanium alloy commonly referred to as Ti grade IV. Titanium exhibits great mechanical properties, resistance to corrosion, and biocompatibility, it has served as a superior biomedical implant material for many years [15].

In order to obtain powerful enough stress waves to initiate delamination, a water-glass layer would need to be added to the back side of our samples. To account for this process a compound substrate assembly, shown in Fig. 18.1b., was developed using a 1″ × 1″ glass slide with reflective coatings on both sides (100 nm thick Ti and 300 nm thick Al). This sample was modified so it could be glued to the bottom of a 35 mm diameter petri-dish. The petri dish had a 13/16″ hole cut in the bottom to expose the titanium to bacteria and growth media. Assemblies were then placed in a spin-coater to apply the water-glass confining layer to the bottom side of the sample. The lower surface that was coated with water-glass was initially isolated with parafilm to protect the surface during incubation of bacteria. However, the surfaces became ununiform and unreliable for stress amplification during incubation as seen in Fig. 18.1c. A new method for preserving the applied water-glass was developed, covering the bottom of the sample assembly with a sacrificial petri dish of the same dimensions and then using stretchable laboratory film to hold the two dishes together. The use of a protective cover greatly enhanced protection of the confining layer and prevented any drying. Controlling these variables is important and in doing so the goal of developing true interfacial stress of a given biofilm and substrate can be quantified. Samples were then sterilized and inoculated with bacteria using Todd Hewitt Yeast broth with 375 mM Sucrose concentration. The sucrose was added to the THY media in order to increase adhesion strength of *S. mutans*. After successful biofilm growth, phosphate buffered saline was applied to preserve bacterial livelihood. Entire assemblies were then mounted into the testing setup in a horizontal orientation. The infrared laser setup now adopted a 45° angled mirror to direct the laser pulse vertically into the bottom of the sample assembly.

Multiple substrate assemblies were tested for each biofilm condition at multiple locations on each biofilm yielding over 100 total loading sites. The onset of failure of the biofilm-titanium interface was marked by spallation of the biofilm from the surface of the titanium resulting in a dark spallation region. For example, Fig. 18.1d, e shows optical images of before and after loading, respectively, at a range of laser fluences. As laser fluence increased, the spallation region also increased.

The graph shown in Fig. 18.1f is includes results from all the samples tested to depict an overall trend of fluence to rate of spallation at a given laser fluence. This process was done to establish a more general view of all the sample data and to clearly present the expected increase of spallation rate with increase of fluence. The fluence values were divided into bins and rate of film failure at those fluences were used. The number of tests at each fluence bin is displayed above each bar. Additional calibration testing must be conducted in order to convert these fluence values to quantitative substrate stresses for the biofilm. In order to do this the mechanical properties of the *S. mutans* biofilm must be determined using AFM techniques. This preliminary work spurred interest in the effect of sucrose on the adhesion of *S. mutans*. Here, we have demonstrated the development of the laser spallation technique to compare the adhesion strength of biofilms on substrates. With the quantitative information obtained an accurate comparison can be made between surface characterizations. The comparisons can be used to construct optimal surface profiles to prevent certain biofilm formations.

Acknowledgements We thank the Center for Pharmaceutical Research and Innovation (CPRI) for use of bacterial culture equipment. CPRI is supported, in part, by the University of Kentucky College of Pharmacy and Center for Clinical and Translational Science (UL1TR001998). We thank Drs. Larissa Ponomareva and Natalia Korotkova for sharing their bacterial culture expertise.

References

1. Kreth, J., Merritt, J., Qi, F.: Bacterial and host interactions of oral streptococci. DNA Cell Biol. **28**(8), 397–403 (2009)
2. Albrektsson, T., et al.: The long-term efficacy of currently used dental implants: a review and proposed criteria of success. Int. J. Oral Maxillofac. Implants. **1**, 11–25 (1986)
3. Steinmann, P.A., Hintermann, H.E.: A review of the mechanical tests for assessment of thin-film adhesion. J. Vac. Sci. Technol. A. **7**(3), 2267–2272 (1989)
4. Mittal, K.L.: Adhesion measurement of thin films. Electrocompon. Sci. Technol. **3**(1), 21–42 (1976)
5. Wang, J., Sottos, N.R., Weaver, R.L.: Mixed-mode failure of thin films using laser-generated shear waves. Exp. Mech. **43**(3), 323–330 (2003)
6. Grady, M.E., et al.: Shockwave loading of mechanochemically active polymer coatings. ACS Appl. Mater. Interfaces. **6**(8), 5350–5355 (2014)
7. Grady, M.E., et al.: Molecular tailoring of interfacial failure. Langmuir. **30**(37), 11096–11102 (2014)
8. Gupta, V., et al.: Measurement of interface strength by a laser spallation technique. J. Mech. Phys. Solids. **40**(1), 141–180 (1992)
9. Wang, J.L., Weaver, R.L., Sottos, N.R.: A parametric study of laser induced thin film spallation. Exp. Mech. **42**(1), 74–83 (2002)
10. Hu, L., et al.: Cell adhesion measurement by laser-induced stress waves. J. Appl. Phys. **100**(8), 084701 (2006)
11. Francis, N.C., et al.: Analysis of flexible substrates for clinical translation of laser-generated shockwave therapy. Biomed. Opt. Express. **6**(3), 827–837 (2015)
12. Miller, P., Hu, L., Wang, J.: Finite element simulation of cell–substrate decohesion by laser-induced stress waves. J. Mech. Behav. Biomed. Mater. **3**(3), 268–277 (2010)
13. Nakamura, H., et al.: Glycosaminoglycan degradation reduces mineralized tissue–titanium interfacial strength. J. Biomed. Mater. Res. A. **77A**(3), 478–486 (2006)
14. Krzyściak, W., et al.: The virulence of Streptococcus mutans and the ability to form biofilms. Eur. J. Clin. Microbiol. Infect. Dis. **33**(4), 499–515 (2014)
15. Niinomi, M.: Mechanical properties of biomedical titanium alloys. Mater. Sci. Eng. A. **243**(1), 231–236 (1998)

Chapter 19
Influence of Adhesion on the Mechanical Response of Granular Composites

Sharada Bhavanam and Nikhil Karanjgaokar

Abstract The adhesion energy between the polymer binder and granular crystals strongly influence the mechanical properties of granular composites under different loading conditions. Thus, the accurate measurement of interfacial adhesion is necessary for the accurate description of the mechanical behavior and failure of granular composites. In the current work, an experimental setup capable of performing a peel test at different displacement rates (10 nm/s–1 mm/s) and peel angles (45°–135°) has been developed to experimentally quantify interfacial adhesion between soft elastomer (binder) and hard substrates (crystals). Since the elastomeric binder in many granular composites materials can be viscoelastic in nature, a time-dependent mathematical framework was adopted to calculate interfacial adhesion for these viscoelastic peel test experiments. Thus, the framework allows for the quantification of interfacial adhesion between different combinations of viscoelastic binders and crystals as a function of peel rate and angle. High-resolution optical microscope and digital image correlation was employed to capture the displacement and strain fields of both linear elastic and viscoelastic material under different conditions.

Keywords Adhesion · Polymer bonded explosive · DIC · Viscoelastic · Peeling

19.1 Introduction

The adhesion at the crystal-binder interface plays an important role on the mechanical behavior of granular composites. Granular composite materials have been employed in various commercial, defense, and space applications such as demolition explosives, combat-safe insensitive munitions, and solid rocket propellants. The mechanical properties of these composites are influenced by (1) properties of the binder and crystal, (2) size, morphology, and arrangement of granular crystals and (3) binder-crystal adhesion. The damage accumulated in these composites is strongly influenced by interfacial debonding. The adhesion at binder-crystal interface also plays an important role on major failure mechanisms such as crystal fracture, shearing of binder, de-bonding of crystal-binder interface, and inter-granular voiding [1]. Literature review has shown that quasi-static and dynamic experiments have been conducted to study the mechanical properties of granular composites under different loading conditions but limited experiments were conducted to study interfacial adhesion of different binders and crystals [2].

Adhesion energy of interfaces in composites has been quantified experimentally through macroscopic and microscopic experiments. Techniques such as Fiber Pull-out test [3], Fiber Push-out test [4], Atomic Force Microscopy [5], JKR (Johnson, Kendall, Roberts) model based contact testing technique [6], Double Cantilever Beam [7], and Peel Test [8] have all been used to quantify adhesion between different materials. These experimental techniques provide information about the interface mechanical properties such as adhesion energies, interfacial strength, and fracture toughness. Since the polymer binders in most granular composites are significantly more compliant than the stiff crystals, the peel test has been employed for the measurement of the adhesion energy of the crystal-binder interface. Kovalchick et al. [9] have developed a macroscale displacement-controlled peel test for obtaining adhesion energy of an elastic adhesive tape from a rigid substrate for the extraction of cohesive zone law from a peel test. Chen et al. [10] have developed a theoretical model for determining the interfacial adhesion of a viscoelastic strip on a rigid substrate. In this paper, we have adopted their viscoelastic framework for determining the adhesion energy of a viscoelastic material. Furthermore, Peng et al. [11] have used this viscoelastic framework to determine the viscoelastic energy release rate of a 3 M Vinyl Electrical Tape thin film from a glass substrate for studying the effect of several parameters such as peeling rate, film thickness, and surface roughness on the peel-off force. In this study, we developed a miniaturized peel testing setup to quantify the effect of peel rate and peel angle on interfacial

S. Bhavanam (✉) · N. Karanjgaokar
Aerospace Engineering Program, Department of Mechanical Engineering, Worcester Polytechnic Institute, Worcester, MA, USA

M. Grady et al. (eds.), *Mechanics of Biological Systems & Micro-and Nanomechanics, Volume 4*, Conference Proceedings of the Society for Experimental Mechanics Series, https://doi.org/10.1007/978-3-319-95062-4_19

adhesion between soft elastomers (i.e., PDMS, HTPB & etc.) and hard substrates (glass, sugar, & Graphene etc.). Digital Image Correlation (DIC) was employed to measure the full field kinematic and strain fields for viscoelastic binder strips during the peel test.

19.2 Experimental Procedure

A custom-built miniaturized peel test capable of performing peel experiments at angles between 45° and 135° and at peel rates of 10 nm/s–1 mm/s was developed to quantify the adhesion between a linear elastic or viscoelastic strip and hard substrate. The peel force during the experiments was measured using a miniaturized load-cell and the peel rate was controlled using a pair of high precision linear actuators with displacement resolution of 20 nm (Fig. 19.1). The preliminary experiments in the current study include peel tests for a linear elastic material (Scotch Double Sided Tape™ 3 mil thick) and a viscoelastic material (PDMS specimens with comparable thickness) mounted on a hard glass substrate. Several different peel angles were tested for tape, ranging from 30° to 90° as can be seen in Figs. 19.2, 19.3, and 19.4. Thick PDMS strips are clamped to the grip whereas thin PDMS strips are usually attached to the grip using 5-min epoxy (Devcon, Danvers, MA). The peel arm length was measured to be 43 mm, the strip width was measured to be 3 mm, and the peel rate was 0.18 mm/min.

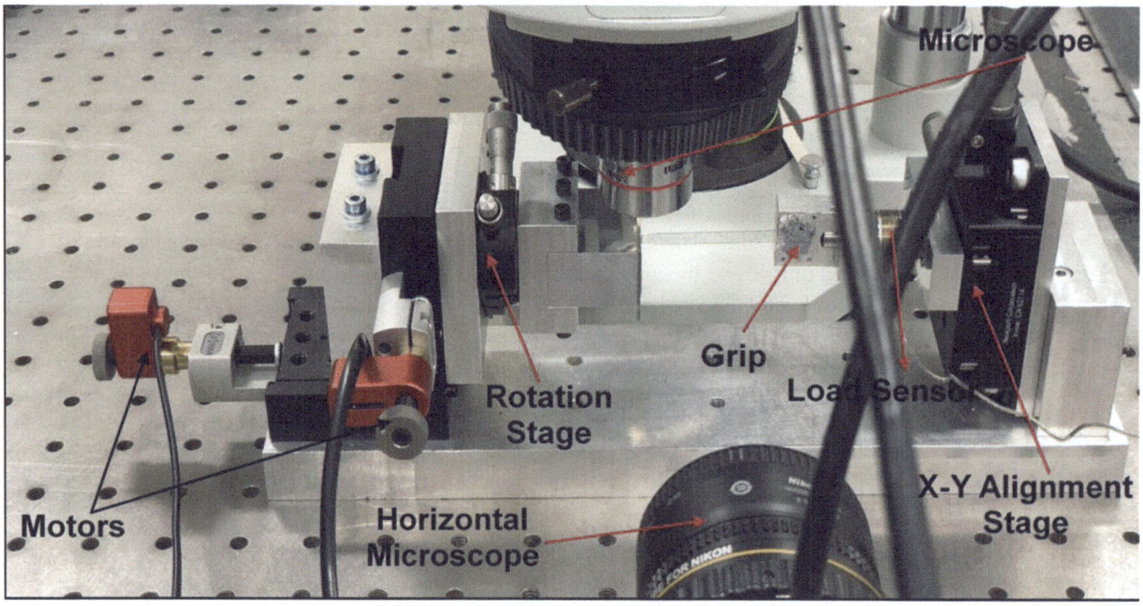

Fig. 19.1 Miniaturized experimental peel test setup designed to measure adhesion energy at microscale. Two motors move in tandem to maintain constant peel angle and peel rate. The setup is placed under an optical microscope for in-situ visualization of peel angle while the horizontal microscope captures the strain field of the specimen

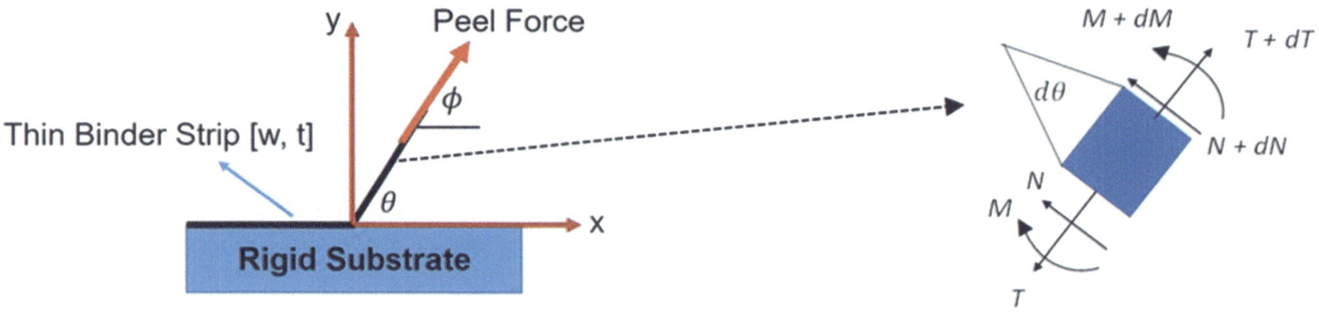

Fig. 19.2 Optical images of linear elastic double-sided tape for different peel angles: (**a**) 30°, (**b**) 45° and (**c**) 90°

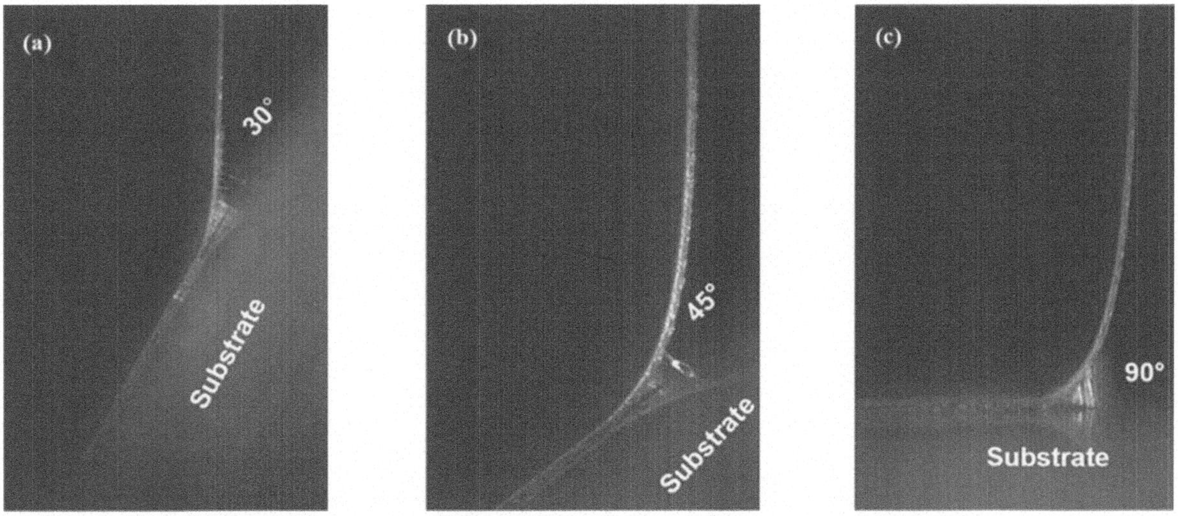

Fig. 19.3 Optical images of the viscoelastic PDMS strips for different peel angles: (**a**) 45° and (**b**) 90°

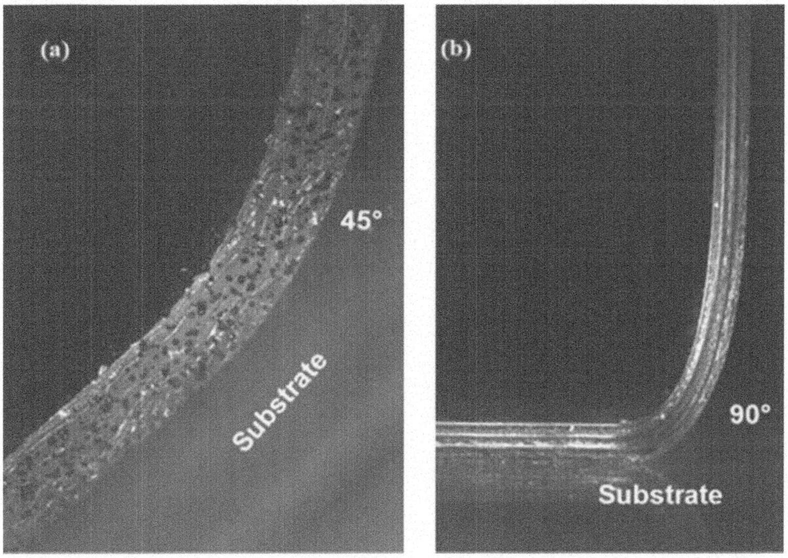

Fig. 19.4 Speckle pattern of the specimen strip mounted in peel testing setup to facilitate the digital image correlation-based displacement measurements during peel tests

In order to compute the adhesion energy of a linear elastic and viscoelastic material, MATLAB (MathWorks, Natick, MA) image processing software was employed to determine the curvature κ for both the linear elastic and viscoelastic strip. Force equilibrium and moment balance equations [10] were used to compute the adhesion energy in terms of the tangential angle θ and peel angle ϕ as shown in Fig. 19.5.

$$\frac{dT}{ds} - KN = 0 \tag{19.1}$$

$$\frac{dN}{ds} + KT = 0 \tag{19.2}$$

$$\frac{dM}{ds} + N = 0 \tag{19.3}$$

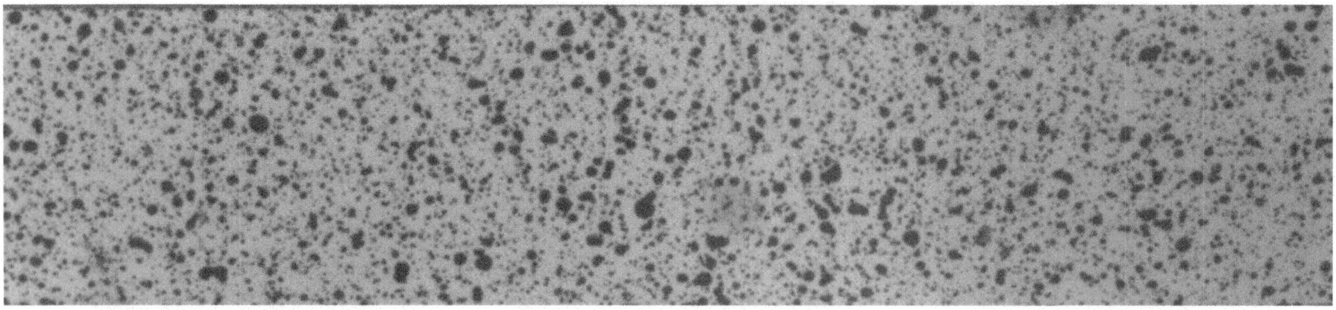

Fig. 19.5 Thin binder strip with thickness t and width w peeled off from a rigid substrate at peel angle ϕ and tangential angle θ, the schematic on the right shows the force/moment balance equilibrium

where T is the axial force, N is the transverse force, K is the curvature, and M is the bending moment along the length of the strip s, by applying the force/moment balance equations, the following expression is obtained for the curvature K of the strip:

$$K = \frac{d\theta}{ds} \tag{19.4}$$

$$K = \frac{2}{h}\left\{6\left(\frac{F}{EH}\right)[1 - \cos(\phi - \theta)]\right\}^{\frac{1}{2}} \tag{19.5}$$

where F is the peel force, E is the elastic modulus of the strip, and H is thickness of the strip. As mentioned earlier, the viscoelastic framework by Chen et al. is adopted in this paper to compute the adhesion energy. Adhesion energy of a viscoelastic material is a function of the bending moment M, axial force T, curvature κ, peel angle ϕ, peel force F, width of strip w, and the membrane strain of the viscoelastic strip ε_m.

$$G = \frac{F}{w}(1 - cos\phi) + \frac{1}{w}\int_0^\infty \kappa \frac{dM}{ds} ds + \frac{1}{w}\int_0^\infty \varepsilon_m \frac{dT}{ds} ds \tag{19.6}$$

For small peeling strain, the above expression can simply be reduced to the following expression:

$$G = \frac{F}{w}(1 - cos\phi) \tag{19.7}$$

For a linear elastic material, Eq. (19.7) can be used to compute the adhesion energy.

19.3 Results and Discussion

The profile of the linear elastic double-sided tape strip at different peel angles of 30°, 45°, and 90° were obtained using an optical microscope based high-resolution camera system as shown in Fig. 19.2. Similarly, the profile of the viscoelastic PDMS strip at peel angles of 45° and 90° are shown in Fig. 19.3. The average adhesion energies of linear elastic material at various peel angles are shown in Fig. 19.6. The average adhesion energies of viscoelastic material at peel angles of 45° and 90° are shown in Fig. 19.7. It can be observed that the average adhesion energy is higher for a linear elastic material when compared to that of a viscoelastic material. The values of adhesion energy were calculated using Eqs. (19.1)–(19.7) for both materials. Digital Image Correlation (DIC) was employed to get the strain fields of the linear elastic and viscoelastic strips and thus calculate the membrane strain at any time instant. Figure 19.8a, b shows the peel force versus displacement for linear elastic and viscoelastic at 45° peel angle respectively. Figure 19.9a, b shows the peel force versus displacement for 90° peel angle for both materials. It can be seen that the peel force for a linear elastic material is larger than for a viscoelastic material as expected. A peel rate used for the peel experiments was 0.18 mm/min.

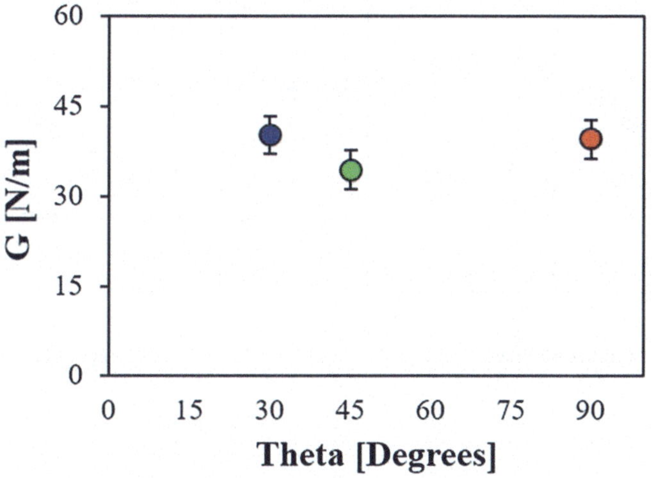

Fig. 19.6 Average adhesion energy of linear elastic double-sided tape mounted on glass substrate for different peel angles

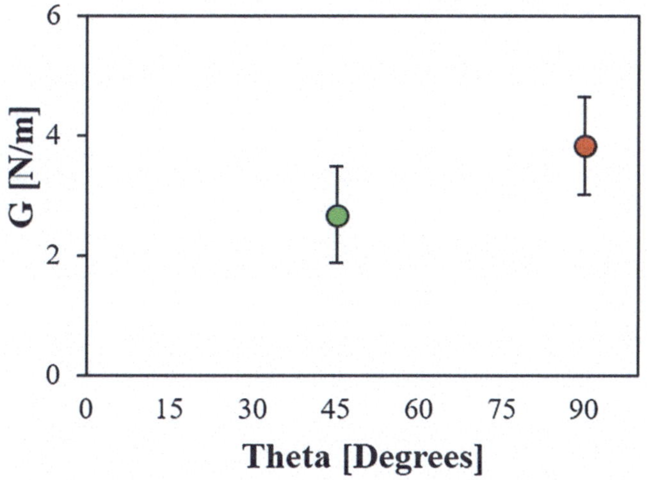

Fig. 19.7 Average adhesion energy of viscoelastic PDMS strips mounted on glass substrate for different peel angles

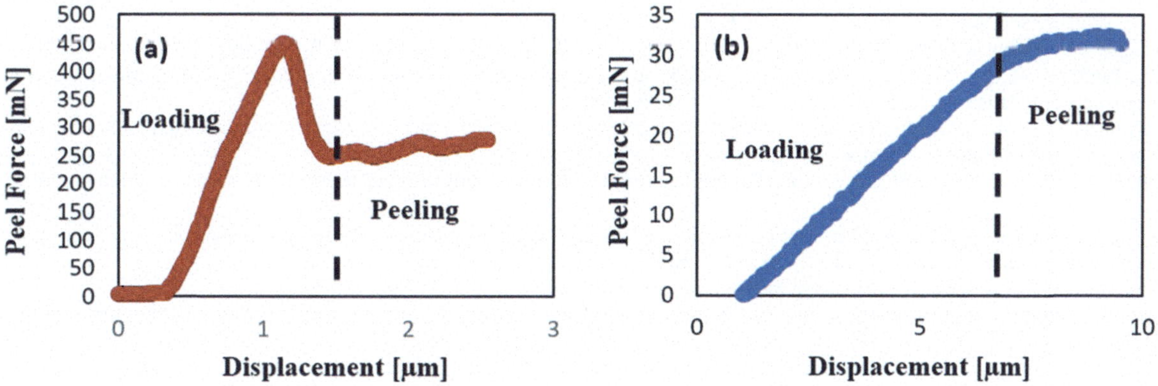

Fig. 19.8 Peel force vs. displacement at 45° peel angle for (**a**) linear elastic double-sided tape, (**b**) viscoelastic PDMS strips mounted on glass substrate

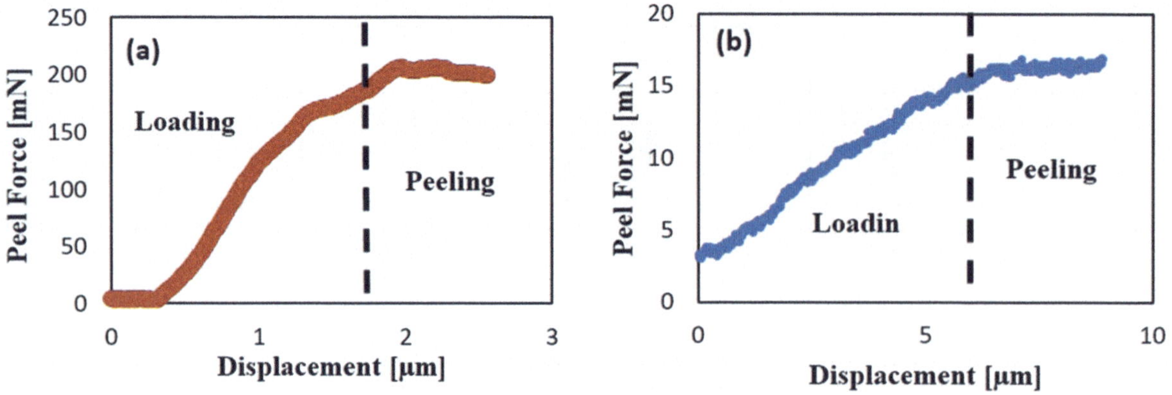

Fig. 19.9 Peel force vs. displacement at 90° peel angle for (**a**) linear elastic double-sided tape, (**b**) viscoelastic PDMS strips mounted on glass substrate

19.4 Conclusions and Future Work

Since interfacial debonding has been established as one of the major failure mechanisms in granular composites, it is critical to understand the role of adhesion on the mechanical response of granular composites. A miniaturized experimental peel test was conducted to quantify adhesion energy of viscoelastic material at the micro-scale. The miniaturized peel test is also capable of performing a peel test for linear elastic materials. Using Digital Image Correlation (DIC), the peel force vs. displacement curves were obtained. The goal is to use the current setup for measuring adhesion energy of various viscoelastic binders adhered to a sugar substrate for understanding the influence of adhesion on the mechanical response of granular composites such as polymer bonded explosives (PBXs). Furthermore, micromechanical models for mock PBX simulants will be developed using (1) cohesive laws for binder-crystal interfaces based on adhesion experiments, (2) DIC based full-field kinematic and strain measurements of mock simulants under quasi-static and dynamic loading, and (3) material models for various binder-crystal combinations.

References

1. Tan, H., Liu, C., Huang, Y., Geubelle, P.H.: The cohesive law for the particle/matrix interfaces in high explosives. J. Mech. Phys. Solids. **53**, 1892–1917 (2005)
2. Siviour, C., Laity, P., Proud, W., et al.: High strain rate properties of a polymer-bonded sugar: their dependence on applied and internal constraints. Proc. R. Soc. A Math. Phys. Eng. Sci. **464**, 1229–1255 (2008)
3. Qiu, Y., Schwartz, P.: Single-fiber pull-out from a microcomposite test. Compos. Sci. Technol. **48**, 5–10 (1993)
4. You, J.H., Lutz, W., Gerger, H., et al.: Fiber push-out study of a copper matrix composite with an engineered interface: experiments and cohesive element simulation. Int. J. Solids Struct. **46**, 4277–4286 (2009)
5. Tsuda, T., Ogasawara, T., Deng, F., Takeda, N.: Direct measurements of interfacial shear strength of multi-walled carbon nanotube/PEEK composite using a nano-pullout method. Compos. Sci. Technol. **71**, 1295–1300 (2011)
6. Charrault, E., Gauthier, C., Marie, P., Schirrer, R.: Experimental and theoretical analysis of a dynamic JKR contact. Langmuir. **25**, 5847–5854 (2009)
7. Yoon, T., Shin, W.C., Kim, T.Y., et al.: Direct measurement of adhesion energy of monolayer graphene as-grown on copper and its application to renewable transfer process. Nano Lett. **12**, 1448–1452 (2012)
8. Kim, K.S., Aravas, N.: Elastoplastic analysis of the peel test. Int. J. Solids Struct. **24**, 417–435 (1988)
9. Kovalchick, C., Molinari, A., Ravichandran, G.: Rate dependent adhesion energy and nonsteady peeling of inextensible tapes. J. Appl. Mech. **81**, 41016 (2013)
10. Chen, H., Feng, X., Huang, Y., et al.: Experiments and viscoelastic analysis of peel test with patterned strips for applications to transfer printing. J. Mech. Phys. Solids. **61**, 1737–1752 (2013)
11. Peng, Z., Wang, C., Chen, L., Chen, S.: Peeling behavior of a viscoelastic thin-film on a rigid substrate. Int. J. Solids Struct. **51**, 4596–4603 (2014)

Sharada Bhavanam B.S. in Aerospace Engineering from University of Illinois at Urbana Champaign. M.S. in Mechanical and Aerospace Engineering from University for California-Irvine. Currently working on my Ph.D. from WPI in Aerospace.

Chapter 20
High-Rate Fracture of Human Skull

C. Allan Gunnarsson, Stephen Alexander, Karin Rafaels, Tim Walter, and Tusit Weerasooriya

Abstract Blunt impacts to the head can transfer energy to the skull and brain, which can cause injuries such as skull fracture, and traumatic brain injury (TBI). The presence of a cranial fracture has been shown to be associated with a higher incidence of intracranial lesions, neurological deficit, and poorer outcome. Therefore, it is important to understand the effect of blunt impact on skull deformation and fracture. In this study, human skull specimens were subjected to blunt loading at high rate in a controlled setting to quantify the deformation, loading, and fracture mechanisms that occur. The microstructure of the specimens were characterized using high-resolution CT (Northstar XRD 1620, at 25 μm resolution). After scanning, the skull specimens were loaded using a high-rate indentation technique. Digital image correlation was used to quantify the skull deformation. Here, the experimental setup and preliminary results are discussed.

Keywords Human skull fracture · High rate material characterization · Digital image correlation · Single loading Kolsky bar

20.1 Introduction

Blunt impacts to the head can transfer energy to the skull and brain, which can cause injuries such as skull fracture, and traumatic brain injury (TBI). The presence of a cranial fracture has consistently been shown to be associated with a higher incidence of intracranial lesions, neurological deficit, and poorer outcome [1]. Therefore, it is important to understand the effect of blunt impact on skull deformation and fracture. In this study, human skull specimens were subjected to blunt loading in a controlled setting to quantify the deformation, loading, and fracture mechanisms that occur. Here, the experimental setup and preliminary results are discussed.

20.2 Experiments

In this study, one human skull cap specimen was created from each of the frontal and left parietal bones of a single post-mortem human subject using a pathology bone saw (EXAKT 512). Care was taken to obtain specimens that did not cross any suture lines. Additionally, the cut line of the specimen was made such that it was perpendicular to the surface normal of the extracted specimen. The two specimens had a geometry similar to a sand dollar, with approximate dimensions of six inch edge diameter and one inch depth. Figure 20.1 shows the donor skull with test specimens removed; the pictured specimen is from the left parietal bone; the frontal specimen in not shown.

The microstructure of the specimens were characterized using high-resolution CT (Northstar XRD 1620, at 25 μm resolution). After scanning, the skull specimens were loaded using a high-rate indentation technique. This indentation was generated via modified Kolsky bar, using an incident bar with a hemispherical indenter tip. The incident bar and indenter tip were made of Al-7075, with a diameter of 31.75 mm (1.25″). The indenter tip was attached to the incident bar using a plastic sleeve, and had a tip curvature radius of 38.1 mm (1.5″). The skull specimens were held to a rigid rectangular backing plate normal to the bar axis and were free to expand radially. The backing plate was mounted to a rigid aluminum frame; the center

C. A. Gunnarsson · K. Rafaels · T. Walter · T. Weerasooriya (✉)
U.S. Army Research Laboratory, Aberdeen Proving Ground, MD, USA
e-mail: carey.a.gunnarsson.civ@mail.mil; tusit.weerasooriya.civ@mail.mil

S. Alexander
SURVICE Engineering Company, Belcamp, MD, USA

M. Grady et al. (eds.), *Mechanics of Biological Systems & Micro-and Nanomechanics, Volume 4*, Conference Proceedings of the Society for Experimental Mechanics Series, https://doi.org/10.1007/978-3-319-95062-4_20

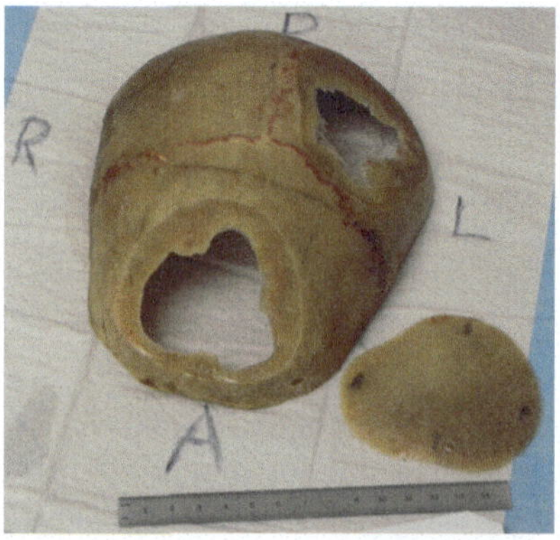

Fig. 20.1 Donor skull with frontal (not shown) and left parietal skull specimens removed. The anterior, posterior, right, and left sides of the skull are denoted by A, P, R and L, respectively, as are the superior (top of the skull) and inferior (bottom, towards the jaw) directions denoted by S and I on the specimen

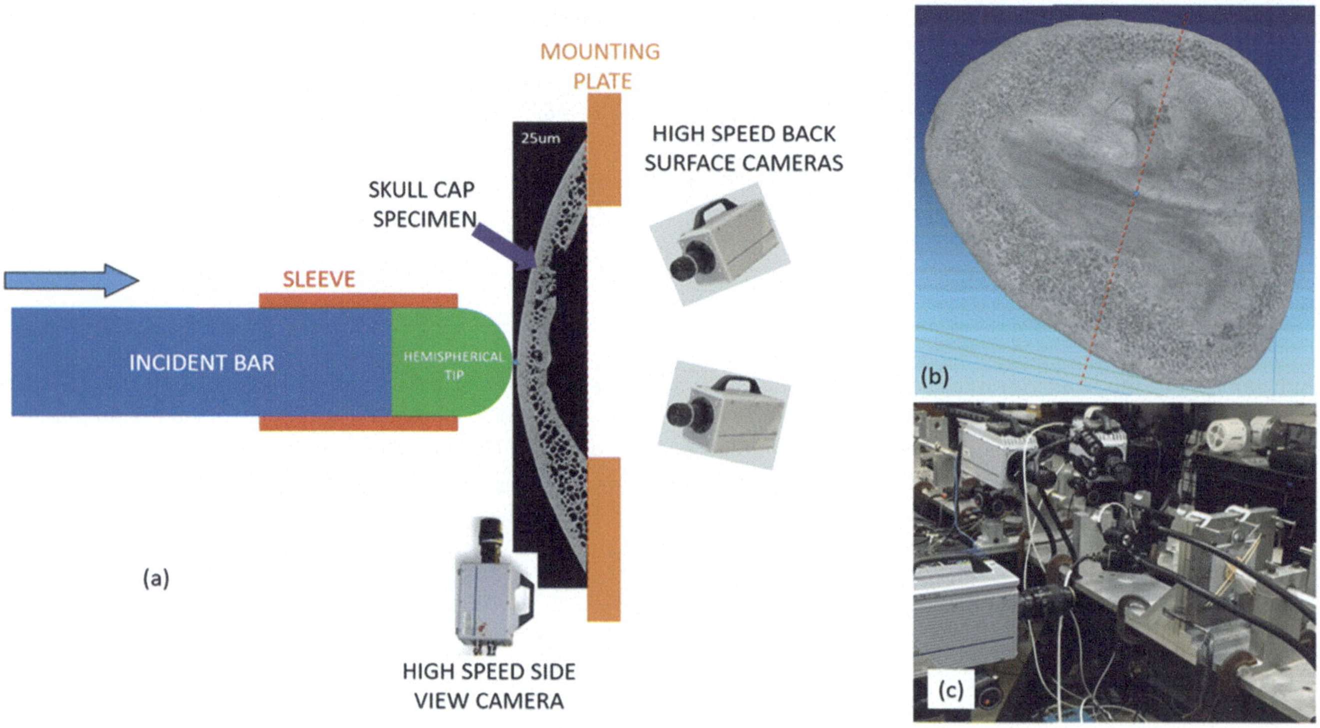

Fig. 20.2 (**a**) Graphical representation of loading configuration, (**b**) 3D CT image of the inner surface of the specimen (here, the left parietal), and (**c**) experimental setup

of the plate was cut to match the size and shape of the specimen's inner raised surface, through which high-speed cameras (Photron SA5 at 50K fps) recorded the skull specimen back surface deformation during loading. The full-field deformation of the inner surface of the skull, opposite the indenter tip, was captured using high-speed digital image correlation (DIC). In addition, a "side view" high-speed camera captured the bar end and indenter tip motion (Photron APX-RS at 50K fps). The experimental setup is graphically depicted in Fig. 20.2a. In the figure, the dashed red line on the 3D inner surface graphic (Fig. 20.2b) shows the location of the slice used in the loading graphic. The light blue dots represent the contact point of the indenter tip with the specimen. The actual experimental setup is shown in Fig. 20.2c. After loading, the skull specimens were CT-scanned to obtain quantitative data on fracture and damage caused to the structure, with the pre-test and post-test images co-registered using DataViewer software (Bruker microCT).

During traditional Kolsky bar experiments, the stress waves travel along the bars many times, which can cause repeated specimen loading. This is generally not a concern for traditional material characterization, as only the first set of pulses are used in analysis to determine the material response. However, here the loading history is being related to damage observed post-experiment. Therefore, a single loading momentum trap setup was utilized to prevent multiple loadings to the skull specimens. This consisted of a flange threaded onto the incident bar at the striker end, and a large sliding mass. A preset gap (created using a micrometer drive) between the flange and mass controlled the indenter tip motion towards the specimen.

During loading, the incident bar and flange moved towards the mass; contact of the flange with the mass prevented any subsequent motion of the indenter tip towards the specimen.

20.3 Results

During loading, the frontal bone skull specimen failed, creating a wedge-shaped fragment which separated from the rest of the specimen. The left parietal bone skull specimen did not show any macro-scale fracture; however, fracture was observed in the specimen micro-structure in the post-experiment CT scan, shown in Fig. 20.3. In the figure, the specimen top view (left) shows the location of two radial cracks, highlighted by the red ellipses. It also indicates the location of the cross-sectional slices for pre-test (top right) and post-test (bottom right) along the dashed green line. The crack in the lower right mark (between P and S) of the top view is highlighted in the post-test cross sectional image.

Typical incident bar strain data, here for the frontal bone specimen, is shown below in Fig. 20.4a. Measurement of the motion of the bar end/indenter tip using traditional strain gage analysis (4.1 m/s) and the side view high-speed camera (3.8 m/s) are both in good agreement, shown in Fig. 20.4b.

The load history at the indenter tip-specimen interface was determined using traditional strain gage analysis; it is shown in Fig. 20.5a (red) for the left parietal bone specimen. The out-of-plane motion of the back surface of the specimen at the impact point (green, measured from the location highlighted in Fig. 20.5b) and indenter tip displacement (blue) are also shown in Fig. 20.5a. The loading rate for the left parietal specimen was approximately 7100 kN/s with a maximum load of 2 kN.

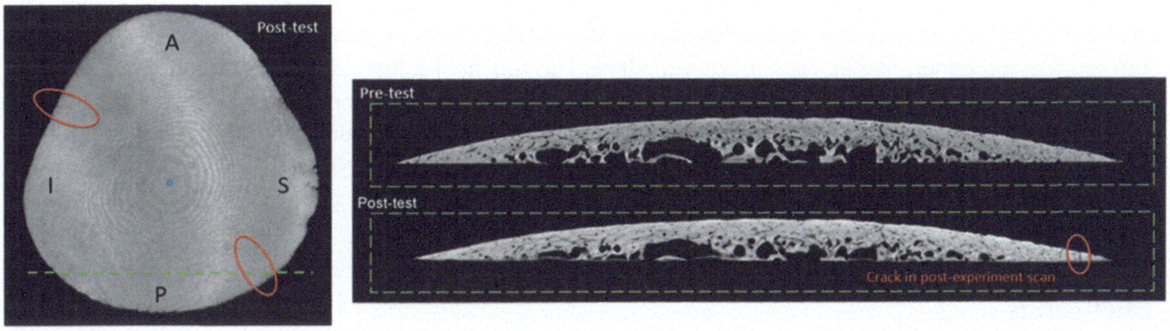

Fig. 20.3 Comparison of pre-test (top right) and post-test (bottom right) CT scan cross-section slices along green dashed line of top view (left) for left parietal specimen

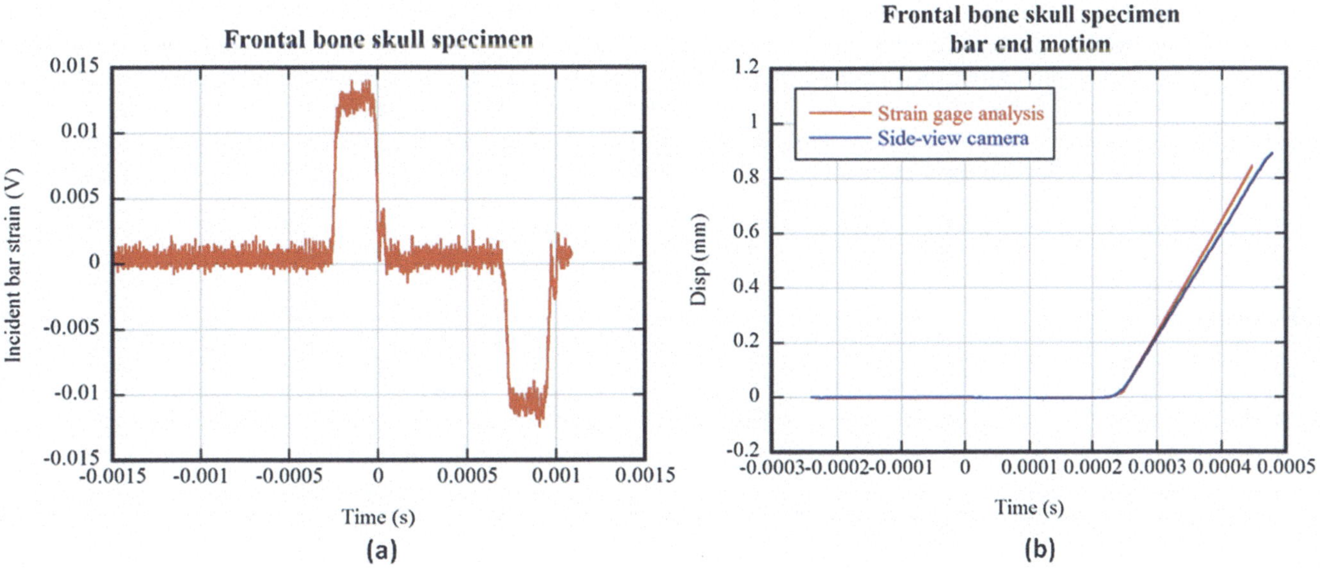

Fig. 20.4 (**a**) Incident and reflected pulses from incident bar strain gage and (**b**) comparison of indenter tip motion from side-view camera and strain gage analysis

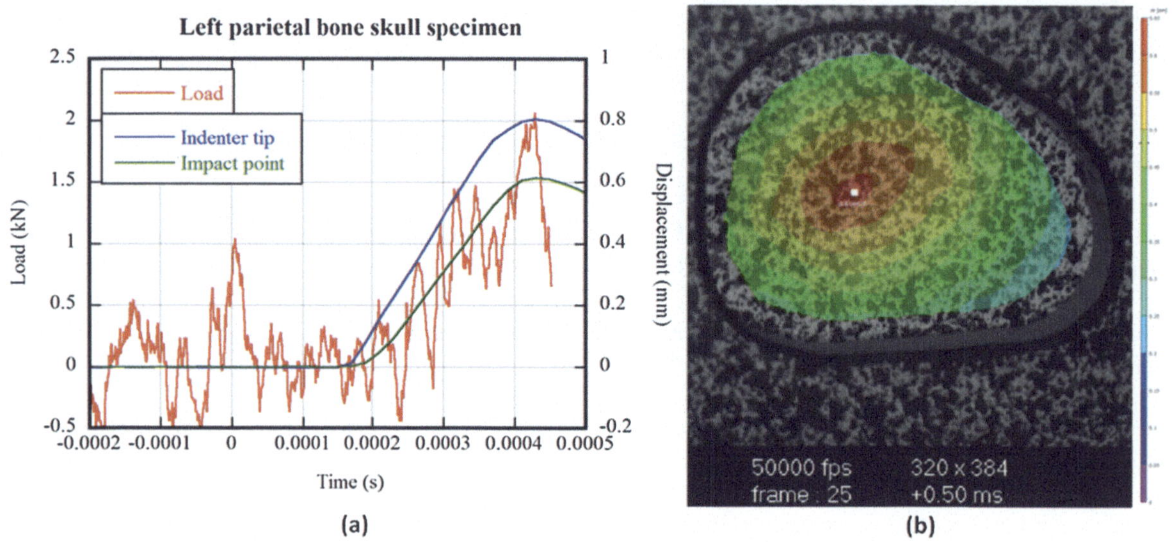

Fig. 20.5 (**a**) Load and displacement history and (**b**) full-field out-of-plane displacement at time of maximum displacement for left parietal skull specimen (scale is 0.0 mm purple to 0.65 mm red)

The indenter tip velocity for the left parietal specimen was measured at 3.7 m/s (camera) and 3.3 m/s (strain gage). The impact point back surface displacement does not match the indenter tip motion; this is presumed to be due to some compression of the skull between the inner and outer surfaces during loading. The full-field out-of-plane displacement for the left parietal specimen at the maximum is shown below in Fig. 20.5b. The location of maximum displacement corresponds to the center of the indenter tip. Figure 20.5b also shows the location (white dot) used to extract the out-of-plane displacement for the impact point during loading. The frontal bone specimen experienced a loading rate of approximately 11,400 kN/s with a maximum load of 4 kN.

20.4 Conclusions

An experimental technique was developed to apply blunt impact loading to the human skull in a controlled environment. This technique was used to load two human skull specimens, one of which exhibited macroscopic fracture, the other demonstrating fracture at the micro-structural level. The loading history and deformation history of both specimens was obtained using an instrumented incident bar and high-speed DIC. The pre-experiment CT scan data will be used to generate a finite element mesh representing the structure of the skull. The FE mesh will be used to simulate these experiments using ARL's human skull model. The experimental full-field displacement data can then be compared to the results from simulation. This comparison of the qualitative skull experimental data (load history, deformation, and fracture) to simulation results will allow for development and evaluation of ARL's human skull model that used to predict the effect of impact on the human skull.

Reference

1. Bullock, M.R., Chesnut, R., Ghajar, J., Gordon, D., Hartl, R., Newell, D.W., Servadei, F., Walters, B.C., Wilberger, J.: Surgical management of depressed cranial fractures. Neurosurgery. **58**(3 Suppl), S2–56–S2–60 (2006). https://doi.org/10.1227/01.NEU.0000210367.14043.0E

C. Allan Gunnarsson has been conducting materials research with the ARL for 10 years. His current research includes development of novel digital image correlation applications to investigate the high rate mechanical behavior of materials.

Chapter 21
Assessment of Fluid Cavitation Threshold Using a Polymeric Split Hopkinson Bar-Confinement Chamber Apparatus

Michael C. Bustamante and Duane S. Cronin

Abstract Mild Traumatic Brain Injury (mTBI) has been associated with blast exposure resulting from the use of improvised explosive devices (IEDs) in recent and past military conflicts. Experimental and numerical models of head blast exposure have demonstrated the potential for high negative pressures occurring within the head at the *contre-coup* location relative to the blast exposure, and it has been hypothesized that this negative pressure could result in cavitation of Cerebrospinal Fluid (CSF) surrounding the brain, leading to brain tissue damage. The cavitation threshold of CSF, the effect of temperature, and the effect of impurities or dissolved gases are presently unknown. In this study, a novel Polymeric Split Hopkinson Pressure Bar and confinement chamber apparatus were used to generate loading in distilled water similar to the conditions in the vicinity of the CSF during blast exposure. Cavitation was identified using high-speed imaging of the event, and a validated numerical model of the apparatus was applied to determine the pressure in the fluid during the exposure. Increasing the water temperature resulted in a decrease in the 50% probability of cavitation from 21 °C (−3320 kPa ± 3%) to 37 °C (−3195 kPa ± 5%) in agreement with the theoretical values, but was not statistically significant. Importantly, the effect of water treatment had a significant effect on the cavitation pressure for water with wetting agent (−3320 kPa ± 3%), degassed water (−1369 kPa ± 16%) and untreated distilled water (−528 kPa ± 25%). Thus, reducing dissolved gases through degassing or the use of a wetting agent significantly increases the cavitation pressure and reduces the variability of the cavitation pressure threshold.

Keywords Fluid cavitation · Polymeric Split Hopkinson Pressure Bar · Mild traumatic brain injury · Negative pressure

21.1 Introduction

Mild Traumatic Brain Injury associated with blast exposure has been a prominent topic of study as a result of the increasing use of IEDs. Brain injury is categorized as mTBI when the individual experiences an alteration or loss of consciousness for up to 30 min [1–3]. There is currently no consensus on the mechanisms that can cause mTBI, with some suggesting: shearing damage of soft-tissue [4–7], distortion of brain cellular structures [5, 8–14], and intracranial fluid cavitation [5, 15–19]. With regard to cavitation, a number of studies have demonstrated that negative pressure can occur within the head at the *contre-coup* location relative to the blast exposure, as a result of pressure wave propagation initiated by a blast wave [16, 19–28]. There is a possibility of cavitation if the negative CSF pressure exceeds the tensile threshold, resulting in the sudden inception, growth, and collapse of a cavitation bubble. Theoretically, the implosive collapse of a near-vacuum bubble results in localized compressive pressures and temperatures several magnitudes above ambient pressure and temperature, respectively [29–31]. CSF cavitation *in vivo* has not been observed and the negative pressure threshold for cavitation is currently unknown. Several numerical studies investigating cavitation due to blast limit the minimum CSF pressure to −100 kPa or the equivalent of 1 atmosphere to simulate the occurrence of cavitation [16, 17, 25, 32]. The aim of this study was to determine the effects of fluid degassing treatment and temperature on the cavitation pressure threshold using distilled water.

M. C. Bustamante (✉) · D. S. Cronin
Department of Mechanical Engineering, University of Waterloo, Waterloo, ON, Canada
e-mail: mcbustam@uwaterloo.ca

M. Grady et al. (eds.), *Mechanics of Biological Systems & Micro-and Nanomechanics, Volume 4*, Conference Proceedings of the Society for Experimental Mechanics Series, https://doi.org/10.1007/978-3-319-95062-4_21

21.2 Methods

Tests with distilled water were performed using a Polymeric Split Hopkinson Pressure Bar (PSHPB) apparatus coupled with a confinement chamber developed for evaluating cavitation (Fig. 21.1). The apparatus comprised a 25.4 mm diameter steel sphere striker, a 25.4 mm diameter and 2.4 m long Polymethyl Methacrylate (PMMA) incident bar, and a 25.4 mm diameter and 61 mm long PMMA confinement chamber. Strain gauges (CEA-13-250UW-120, Micro-Measurements) were mounted at the midpoint of the incident bar. Measured strains were amplified (2210B Signal Conditioning Amplifier, Vishay) and recorded (BNC-2110, National Instruments) at 2 MHz. A Photon Doppler Velocimeter (PDV) (1550 nm wavelength, Ohio Manufacturing Institute) probe was directed at the free-end of the chamber, measuring surface velocity at a recording frequency of 10 MHz. A high-speed camera (FASTCAM SA5 Model 1300K-M1, Photron, 50,000 fps with 512 × 272 resolution) was oriented perpendicular to the transparent chamber, and used to determine the occurrence of cavitation [33–36].

The apparatus generated an incident pressure pulse comparable to that observed propagating in the brain originating from the *coup* of blast exposure [16, 19–28, 34]. The incident pulse propagated along the incident bar, through the chamber and fluid, and reflected in tension at the free-end of the chamber, generating negative fluid pressure localized adjacent to the fluid-chamber interface. It was concluded in a previous study that the implementation of a pressure transducer resulted in reduced cavitation thresholds due to discontinuous geometries and large impedance mismatches [34]. Therefore, a numerical model was developed to predict the apparatus and fluid dynamics. The axisymmetric numerical model consisted of 203,000 axisymmetric quadrilateral elements (~0.4 mm), following a mesh refinement study. A previous study by the authors describes the apparatus and numerical model in full detail and demonstrates the efficacy of the model in predicting strain wave propagation, surface velocities, and fluid pressure changes [34, 36].

Experimental cavitation threshold tests were performed with increasing striker velocities. The tests started at sub-cavitation loading, confirmed through high-speed imaging, and increased until cavitation was observed. After cavitation, the water was discarded, a new water sample was prepared, and the process of increasing striker velocities was restarted. The experimental striker velocities were used with the numerical model to predict the negative pressure occurring during the experiment. Lastly, a sigmoid-shaped probability curve was calculated with the procedure outlined in ISO/TS 18506 using the dataset of cavitation occurrence versus the peak predicted negative water pressure. The pressure corresponding to the 50% probability of cavitation was identified as the cavitation pressure threshold [37].

Three different water conditions (wetting agent, degassed, and non-degassed) and two different temperatures (room temperature, 21 °C, and body temperature, 37 °C) were tested. Non-degassed samples were created by using distilled water as-is without any other treatments. For tests with a wetting agent, a thin film of ~1% concentration wetting agent (Jet–Dry Rinse Agent, Finish) and distilled water solution was applied to the inner-walls of the chamber before filling with non-degassed water. For the degassed condition, an open-chamber filled with water was degassed with an ambient pressure of −93 kPa for at least 20 min. The ambient pressure was slowly decreased to avoid vigorous agitation of the water. Samples requiring room temperature were left to sit at room temperature for several hours before testing. Samples requiring body temperature were prepared by placing a sealed-chamber filled with water into a 37 °C water bath and verifying the test sample temperature with an infrared thermometer immediately before testing.

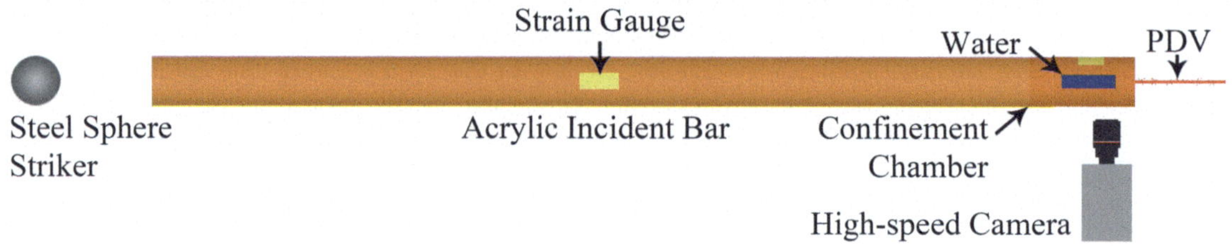

Fig. 21.1 Polymeric Split Hopkinson Pressure Bar with confinement chamber apparatus for measuring fluid cavitation threshold and instrumentation

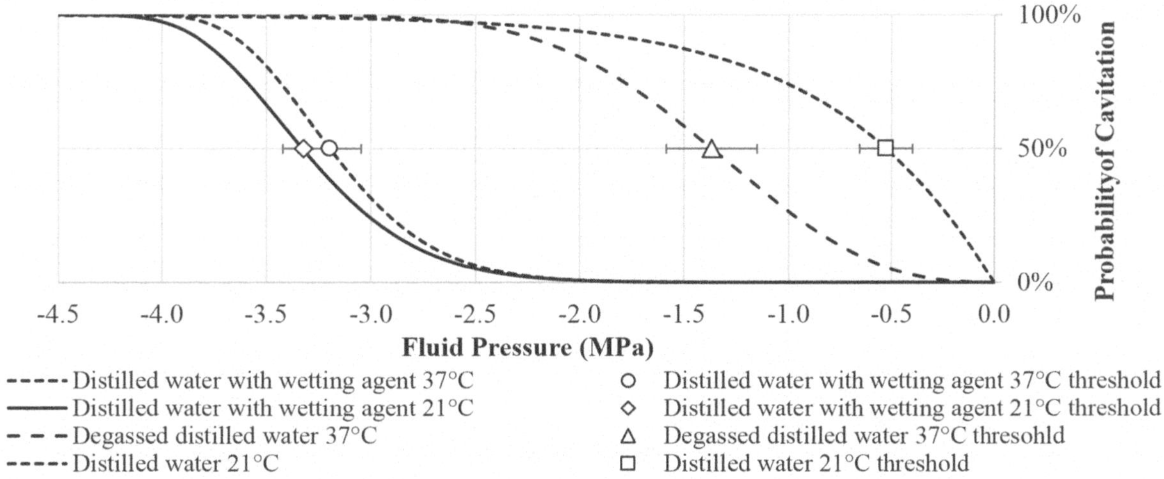

Fig. 21.2 Cavitation probability curves for distilled water including the effects of temperature and degassing method (error bars show one standard deviation)

21.3 Results and Discussion

Fifteen tests were performed with distilled water at 37 °C with a wetting agent applied to the inner-walls of the chamber to minimize surface tension, and therefore minimize entrapped air. A cavitation pressure threshold of −3195 kPa ± 5% (Fig. 21.2) was calculated (ISO/TS 18506), demonstrating an expected decrease in threshold relative to −3320 kPa ± 3% measured in a previous study for distilled water at 21 °C with the same wetting agent (Fig. 21.2) [34]. The measured 4% threshold decrease was comparable to a predicted 3% decrease calculated with the assumption that surface tension is the driving mechanism resisting cavity growth [30, 38, 39]. It should be noted that when considering the variability in the data, the average threshold decrease was not statistically significant.

Thirty-nine cavitation tests were performed with degassed distilled water at 37 °C without a wetting agent, resulting in a calculated threshold of −1369 kPa ± 16%. Similarly, 39 tests were performed with non-degassed distilled water at 21 °C without a wetting agent, resulting in a calculated threshold of −528 kPa ± 25%. Lower variability and a higher cavitation threshold were observed in the degassed water relative to the non-degassed water. Both observations were attributed to the reduction of entrapped air at the water-chamber interfaces, resulting in a higher tolerance for tension. There is further support for this conclusion when both thresholds were compared to non-degassed distilled water with a wetting agent (−3320 kPa ± 3%), which demonstrated a much higher threshold and lower variability [34]. The effect of the wetting agent can be seen in the 21 °C water tests, yielding an increase of approximately 630% in negative threshold relative to non-degassed conditions. With the assumption that the 37 °C water tests exhibit similar effects as the 21 °C tests, albeit lesser negative thresholds, degassing yields an increase of approximately 270% in negative threshold relative to non-degassed conditions.

21.4 Conclusions

Cavitation tests were undertaken for distilled water with varying temperature and pretreatment using a PSHPB coupled to a confinement chamber, high-speed imaging to identify cavitation, and a computational model to determine the fluid pressure during the exposure. It was determined that increasing the temperature of the distilled water with a wetting agent from 21 to 37 °C resulted in a small decrease in cavitation threshold, in agreement with theoretical calculations albeit not statistically significant. Pretreatments to the distilled water including degassing and the use of a wetting agent had the largest effect in terms of increasing the cavitation threshold while also reducing variability in the experimental data. It is concluded that the preparation of fluid samples for cavitation tests can have a large effect on the measured cavitation threshold.

Acknowledgements The authors would like to acknowledge the Natural Sciences and Engineering Research Council of Canada for financial support, and Compute Canada and Sharcnet for providing the necessary computing resources.

References

1. Ruff, R.: Two decades of advances in understanding of mild traumatic brain injury. J. Head Trauma Rehabil. **20**, 5–18 (2005). https://doi.org/10.1097/00001199-200501000-00003
2. McCrea, M.: Mild Traumatic Brain Injury and Postconcussion Syndrome: The New York Evidence Base for Diagnosis and Treatment. Oxford University Press, New York (2008)
3. Gardener, R.C., Yaffe, K.: Epidemiology of mild traumatic brain injury and neurodegenerative disease Raquel. Mol. Cell. Neurosci. **66**, 75–80 (2015). https://doi.org/10.1016/j.mcn.2015.03.001.Epidemiology
4. Alley, M.D., Schimizze, B.R., Son, S.F.: Experimental modeling of explosive blast-related traumatic brain injuries. Neuroimage. **54**, S45–S54 (2011). https://doi.org/10.1016/j.neuroimage.2010.05.030
5. Sarntinoranont, M., Lee, S.J., Hong, Y., et al.: High-strain-rate brain injury model using submerged acute rat brain tissue slices. J. Neurotrauma. **29**, 418–429 (2012). https://doi.org/10.1089/neu.2011.1772
6. Nie, X., Sanborn, B., Weerasooriya, T., Chen, W.: High-rate bulk and shear responses of bovine brain tissue. Int J Impact Eng. **53**, 56–61 (2013). https://doi.org/10.1016/j.ijimpeng.2012.07.012
7. Sosa, M.A.G., De Gasperi, R., Paulino, A.J., et al.: Blast overpressure induces shear-related injuries in the brain of rats exposed to a mild traumatic brain injury. Acta Neuropathol. Commun. **1**, 51 (2013). https://doi.org/10.1186/2051-5960-1-51
8. Ling, G., Bandak, F., Armonda, R., et al.: Explosive blast neurotrauma. J. Neurotrauma. **26**, 815–825 (2009). https://doi.org/10.1089/neu.2007.0484
9. Bo, C., Balzer, J., Brown, K.A., et al.: Development of a chamber to investigate high-intensity compression waves upon live cell cultures. Eur. Phys. J. Appl. Phys. **55**, 31201 (2011). https://doi.org/10.1051/epjap/2011110052
10. Risling, M., Plantman, S., Angeria, M., et al.: Mechanisms of blast induced brain injuries, experimental studies in rats. Neuroimage. **54**, S89–S97 (2011). https://doi.org/10.1016/j.neuroimage.2010.05.031
11. Bolander, R., Mathie, B., Bir, C., et al.: Skull flexure as a contributing factor in the mechanism of injury in the rat when exposed to a shock wave. Ann. Biomed. Eng. **39**, 2550–2559 (2011). https://doi.org/10.1007/s10439-011-0343-0
12. Ryu, J., Horkayne-Szakaly, I., Xu, L., et al.: The problem of axonal injury in the brains of veterans with histories of blast exposure. Acta Neuropathol. Commun. **2**, 153 (2014). https://doi.org/10.1186/s40478-014-0153-3
13. Heldt, S.A., Elberger, A.J., Deng, Y., et al.: A novel closed-head model of mild traumatic brain injury caused by primary overpressure blast to the cranium produces sustained emotional deficits in mice. Front Neurol. **5**, 1–14 (2014). https://doi.org/10.3389/fneur.2014.00002
14. Kamnaksh, A., Budde, M.D., Kovesdi, E., et al.: Diffusion tensor imaging reveals acute subcortical changes after mild blast-induced traumatic brain injury. Sci. Rep. **4**, 4809 (2014). https://doi.org/10.1038/srep04809
15. Lee, C.S., Frizzell, L.A.: Exposure levels for ultrasonic cavitation in the mouse neonate. Ultrasound Med. Biol. **14**, 735–742 (1988). https://doi.org/10.1016/0301-5629(88)90029-4
16. Goeller, J., Wardlaw, A., Treichler, D., et al.: Investigation of cavitation as a possible damage mechanism in blast-induced traumatic brain injury. J. Neurotrauma. **29**, 1970–1981 (2012). https://doi.org/10.1089/neu.2011.2224
17. Panzer, M.B., Myers, B.S., Capehart, B.P., Bass, C.R.: Development of a finite element model for blast brain injury and the effects of CSF cavitation. Ann. Biomed. Eng. **40**, 1530–1544 (2012). https://doi.org/10.1007/s10439-012-0519-2
18. Hong, Y., Sarntinoranont, M., Subhash, G., et al.: Localized tissue surrogate deformation due to controlled single bubble cavitation. Exp. Mech. **56**(1), 97–109 (2015). https://doi.org/10.1007/s11340-015-0024-2
19. Singh, D., Cronin, D.S., Haladuick, T.N.: Head and brain response to blast using sagittal and transverse finite element models. Int J Numer Method Biomed Eng. **30**, 470–489 (2014). https://doi.org/10.1002/cnm.2612
20. Bir, C.: Measuring Blast-Related Intracranial Pressure Within the Human Head. Detroit (2011)
21. Ganpule, S., Alai, A., Plougonven, E., Chandra, N.: Mechanics of blast loading on the head models in the study of traumatic brain injury using experimental and computational approaches. Biomech. Model. Mechanobiol. **12**, 511–531 (2013). https://doi.org/10.1007/s10237-012-0421-8
22. Grujicic, M., Arakere, G., He, T.: Material-modeling and structural-mechanics aspects of the traumatic brain injury problem. Multidiscip. Model. Mater. Struct. **6**, 335–363 (2010). https://doi.org/10.1108/15736101011080097
23. Hua, Y., Kumar Akula, P., Gu, L., et al.: Experimental and numerical investigation of the mechanism of blast wave transmission through a surrogate head. J. Comput. Nonlinear Dyn. **9**, 31010 (2014). https://doi.org/10.1115/1.4026156
24. Moss, W.C., King, M.J., Blackman, E.G.: Skull flexure from blast waves: a mechanism for brain injury with implications for helmet design. Phys. Rev. Lett. **103**, 4–7 (2009). https://doi.org/10.1103/PhysRevLett.103.108702
25. Panzer, M.B., Bass, C.R., Myers, B.S.: Numerical Study on the Role of Helmet Protection in Blast Brain Injury. In: Pers. Armor Syst. Symp. Quebec City, p. 11 (2010)
26. El Sayed, T., Mota, A., Fraternali, F., Ortiz, M.: Biomechanics of traumatic brain injury. Comput. Methods Appl. Mech. Eng. **197**, 4692–4701 (2008). https://doi.org/10.1016/j.cma.2008.06.006
27. Zhang, L., Makwana, R., Sharma, S.: Brain response to primary blast wave using validated finite element models of human head and advanced combat helmet. Front Neurol. **4**, 88 (2013). https://doi.org/10.3389/fneur.2013.00088
28. Zhu, F., Mao, H., Dal Cengio Leonardi, A., et al.: Development of an FE model of the rat head subjected to air shock loading. Stapp Car Crash J. **54**, 211–225 (2010). 2010-22-0011 [pii]
29. Thiruvengadam, A.: Handbook of Cavitation Erosion. Hydronautics, Inc., Laurel (1974)
30. Brennen, C.E.: Cavitation and bubble dynamics. Annu. Rev. Fluid Mech. (1995). https://doi.org/10.1017/CBO9781107338760
31. Knapp, R.T., Daily, J.W., Hammitt, F.G.: Cavitation. McGraw-Hill, New York (1970)
32. Wardlaw, A., Goeller, J.: Cavitation as a possible Traumatic Brain Injury (TBI) damage mechanism. IFMBE Proc. **32**, 34–37 (2010). https://doi.org/10.1007/978-3-642-14998-6_9
33. Salisbury, C., Cronin, D., Lien, F.-S.: Deformation mechanics of a non-linear hyper-viscoelastic porous material, part I: testing and constitutive modeling of non-porous polychloroprene. J Dyn Behav Mater. **1**, 237–248 (2015). https://doi.org/10.1007/s40870-015-0026-2

34. Bustamante, M.C., Singh, D., Cronin, D.S.: Polymeric Hopkinson bar-confinement chamber apparatus to evaluate fluid cavitation. Exp. Mech. **58**(1), 55–74 (2018). https://doi.org/10.1007/s11340-017-0323-x
35. Bustamante, M., Singh, D., Cronin, D.S.: Modified hopkinson apparatus to investigate fluid cavitation as a potential source of injury. Conf Proc Soc Exp Mech Ser. **85**, 43–51 (2016). https://doi.org/10.1007/978-3-319-22452-7_8
36. Bustamante, M., Cronin, D.S., Singh, D.: Experimental testing and computational analysis of viscoelastic wave propagation in polymeric split Hopkinson pressure bar. In: Kimberley, J., Lamberson, L., Mates, S. (eds.) Dynamic Behavior of Materials, Volume 1. Conference Proceedings of the Society for Experimental Mechanics Series, pp. 67–72. Springer, Cham (2018)
37. International Organization for Standardization: Procedure to Construct Injury Risk Curves for the Evaluation of Road User Protection in Crash Tests (ISO/TS 18506), 1st edn. International Organization for Standardization (ISO), Geneva (2014)
38. Vargaftik, N.B., Volkov, B.N., Voljak, L.D.: International tables of the surface tension of water. J. Phys. Chem. Ref. Data. **12**, 817–820 (1983). https://doi.org/10.1063/1.555688
39. Kayser, W.V.: Temperature dependence of the surface tension of water in contact with its saturated vapor. J. Colloid Interface Sci. **56**, 622–627 (1976). https://doi.org/10.1016/0021-9797(76)90130-2

Michael C. Bustamante A University of Waterloo MASc graduate in Mechanical Engineering. My main research topic for my MASc was in CSF cavitation as a result of blast exposure. I am currently a Research Associate in the University of Waterloo as a modeler experimentalist.

Chapter 22
Contact Reliability of Pt- and TiN-Coated Microswitches in Different Environments

Changho Oh and Maarten P. de Boer

Keywords MEMS switches · Contact materials · Gas environments · Hydrocarbons · Reliability

Ohmic micro- and nanoswitches are of interest for a wide variety of applications including phased array radars and as low power complements to transistors [1–4]. The ohmic switches prevent any leakage current when the device is in the OFF state by forming an air gap that separates the electrical contacts. The subthreshold swing can also be effectively reduced to less than 1 mV/decade [5], which can further reduce the power consumption. Ohmic nanoswitches can potentially operate at ∼20 mV and will consume up to $\sim 10^6$ times less power than MOSFETs. However, the commercialization of these mechanical switches has been hindered due to a number of reliability issues, especially on switch contacts.

Past MEMS switch reliability studies suggest that tribopolymer (TP) forms to varying degrees [6], but few controlled studies that pinpoint the failure mechanisms and identify the best contact materials and environments have been conducted. Recent studies have clearly identified hydrocarbon (HC) contamination as a major source of degradation [7, 8]. It leads to the formation of TP, which increases electrical contact resistance (ECR). Additionally, the gas environment may also be responsible for mechanochemical effects, which could affect ECR [7–9].

Pt is a frequently-used material in macroscale switches [10] and has often been used in MEMS switches [6–9, 11, 12] because of its good hardness, resistance to oxidation and high electrical conductivity. However, due to its catalytic properties, it is more susceptible to TP formation. TiN is of interest due to its high hardness and chemical inertness [13–16]. Here, the performance of Pt and TiN contact materials were compared in different environments. The microswitches were coated in the Carnegie Mellon University's nanofabrication facility (Fig. 22.1). They were loaded into our test system, and surface cleaning was performed via an *in-situ* high-temperature bake-out under a high vacuum condition or an *in-situ* oxygen plasma cleaning (Fig. 22.2) to obtain an initially clean surface by removing adventitious hydrocarbons [17, 18]. This *in-situ* procedure prevents re-contamination, which can otherwise occur by exposing cleaned samples to air. Subsequently, either clean N2, clean O2 or N_2 with HC contaminant gas was introduced into the chamber and switches were cycled up to 10^8 times while monitoring ECR. Switch contact surfaces were characterized using Scanning Electron Spectroscopy (SEM) before and after cycling.

Pt-coated microswitches were tested with or without prior surface cleaning in a clean nitrogen gas environment. Plasma-cleaned switches demonstrated six orders of magnitude greater cycles to failure, as measured by a change in ECR [18]. This is because the plasma eliminated adventitious HCs, as indicated by a linear *I-V* characteristic across the contacts at a bias as low as 1 mV.

TiN was tested to determine if its chemical inertness remains intact under high stress conditions. Prior to testing, TiN-coated switches are cleaned with an *in-situ* ultra-high vacuum bakeout. In N_2 and Ar gases, the TiN-coated switches failed prematurely by mechanical damage and failed even more quickly in O_2 gas. In strong contrast, they performed better in the HC-rich environment.

We will report details of both chemical and mechanical effects in each system. This work adds useful information to the goal of developing guidelines to select combinations of contact materials and environments that effectively resist TP formation and improve MEMS switch reliability.

C. Oh · M. P. de Boer (✉)
Department of Mechanical Engineering, Carnegie Mellon University, Pittsburgh, PA, USA
e-mail: changhoo@andrew.cmu.edu; mpdebo@andrew.cmu.edu

M. Grady et al. (eds.), *Mechanics of Biological Systems & Micro-and Nanomechanics, Volume 4*, Conference Proceedings of the Society for Experimental Mechanics Series, https://doi.org/10.1007/978-3-319-95062-4_22

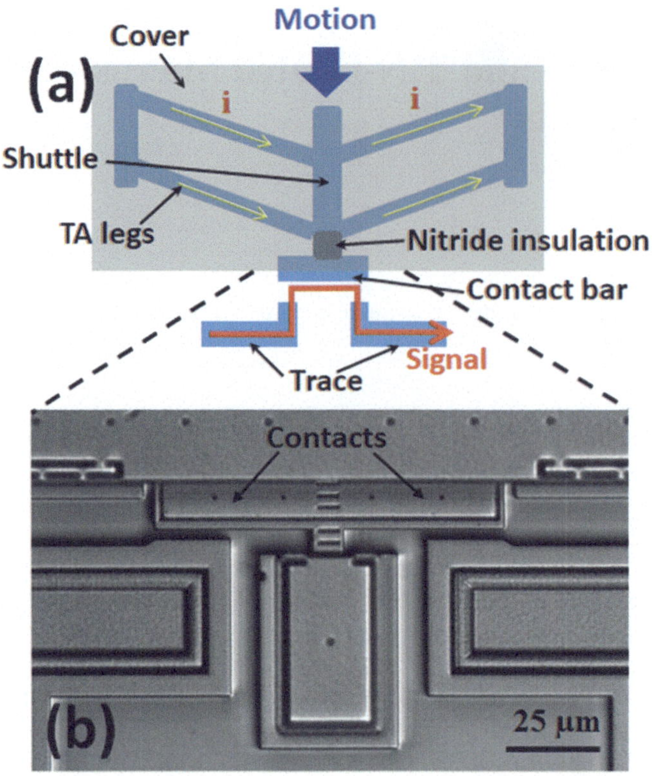

Fig. 22.1 (**a**) Schematic of thermally-actuated MEM switch, (**b**) an optical image of the switch contacts. The contact bar moves down to close the switch

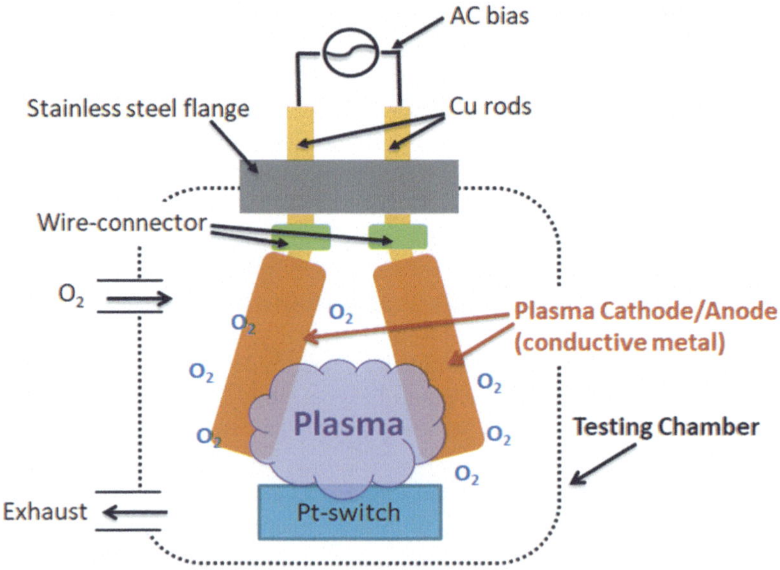

Fig. 22.2 Schematic of the in-situ plasma cleaning system

References

1. Spencer, M., et al.: Demonstration of integrated micro-electro-mechanical relay circuits for VLSI applications. IEEE J. Solid State Circuits. **46**(1), 308–320 (2011)
2. Pott, V., Kam, H., Nathanael, R., Jeon, J., Alon, E., King Liu, T.J.: Mechanical computing redux: relays for integrated circuit applications. Proc. IEEE. **98**(12), 2076–2094 (2010)
3. Lee, J.O., et al.: A sub-1-volt nanoelectromechanical switching device. Nat. Nanotechnol. **8**(1), 36–40 (2012)

4. Loh, O.Y., Espinosa, H.D.: Nanoelectromechanical contact switches. Nat. Nanotechnol. **7**(5), 283–295 (2012)
5. Sinha, N., Guo, Z., Felmetsger, V., Piazza, G.: 100 NM Thick Aluminum Nitride Based Piezoelectric Nano Switches Exhibiting 1 MV Threshold Voltage via Body-Biasing, no. Cicc (2010)
6. Czaplewski, D.A., Nordquist, C.D., Dyck, C.W., Patrizi, G.A., Kraus, G.M., Cowan, W.D.: Lifetime limitations of ohmic, contacting RF MEMS switches with Au, Pt and Ir contact materials due to accumulation of 'friction polymer' on the contacts. J. Micromech. Microeng. **22**(10), 105005 (2012)
7. Brand, V., Baker, M.S., de Boer, M.P.: Impact of contact materials and operating conditions on stability of micromechanical switches. Tribol. Lett. **51**, 341–356 (2013)
8. Brand, V., Baker, M.S., de Boer, M.P.: Contamination thresholds of Pt- and RuO2-coated ohmic switches. J. Microelectromech. Syst. **22**(6), 1248–1250 (2013)
9. de Boer, M.P., Czaplewski, D.A., Baker, M.S., Wolfley, S.L., Ohlhausen, J.A.: Design, fabrication, performance and reliability of Pt- and RuO2-coated microrelays tested in ultra-high purity gas environments. J. Micromech. Microeng. **22**(10), 105027 (2012)
10. ASM Handbook Volume 2: Properties and Selection: Nonferrous Alloys and Special-Purpose Materials, 10th ed. ASM International (1990)
11. Lee, H., Coutu, R.A., Mall, S., Leedy, K.D.: Characterization of metal and metal alloy films as contact materials in MEMS switches. J. Micromech. Microeng. **16**(3), 557–563 (2006)
12. Chen, L., et al.: Contact resistance study of noble metals and alloy films using a scanning probe microscope test station. J. Appl. Phys. **102**(7), 74910 (2007)
13. Jang, W.W., et al.: Fabrication and characterization of a nanoelectromechanical switch with 15-nm-thick suspension air gap. Appl. Phys. Lett. **92**(10), 103110 (2008)
14. Jang, W.W., et al.: NEMS switch with 30nm-thick beam and 20nm-thick air-gap for high density non-volatile memory applications. Solid State Electron. **52**(10), 1578–1583 (2008)
15. Ramezani, M., Severi, S., Moussa, A., Osman, H., Tilmans, H.A.C., De Meyer, K.: Contact reliability improvement of a poly-sige based nano-relay with titanium nitride coating. **1**, 576–579 (2015)
16. Seneviratne, D., Nielson, G.N., Takahashi, S., Barbastathis, G., Tuller, H.L.: On the use of titanium nitride as structural material for nano-electro-mechanical systems (NEMS). In: 5th IEEE Conference on Nanotechnology, pp. 292–295 (2005)
17. Oh, C., Streller, F., Carpick, R.W., de Boer, M.P.: Effectiveness of oxygen plasma versus UHV bakeout in cleaning MEMS switch surfaces. In: 2015 IEEE 61st Holm Conference on Electrical Contacts, pp. 358–362 (2015)
18. Oh, C., Streller, F., Ashurst, W.R., Carpick, R.W., de Boer, M.P.: In-situ oxygen plasma cleaning of microswitch surfaces—comparison of Ti and graphite electrodes. J. Micromech. Microeng. **26**(11), 115020 (2016)

Changho Oh a Mechanical Engineering PhD candidate at Carnegie Mellon University. My research focus is to improve MEMS switch reliability by studying the effects of (1) device materials and (2) surrounding gas environments on the actual device performance.

The manufacturer's authorised representative in the EU is Springer Nature Customer Service Centre GmbH, Europaplatz 3, 69115 Heidelberg, Germany. If you have any concerns regarding our products, please contact ProductSafety@springernature.com

Printed and bound by CPI Group (UK) Ltd, Croydon, CR0 4YY
06/10/2026
02228820-0002